Mindsteps to the Cosmos

K&E Series on Knots and Everything — Vol. 31

Mindsteps to the Cosmos

Gerald S. Hawkins

Former Chairman, Astronomy Department
Boston University, USA

Published by

World Scientific Publishing Co. Pte. Ltd.
P O Box 128, Farrer Road, Singapore 912805
USA office: Suite 1B, 1060 Main Street, River Edge, NJ 07661
UK office: 57 Shelton Street, Covent Garden, London WC2H 9HE

British Library Cataloguing-in-Publication Data
A catalogue record for this book is available from the British Library.

First published in 1983 by Harper & Row, Publishers, Inc.

MINDSTEPS TO THE COSMOS

ISBN 981-238-123-6

This book is printed on acid-free paper.

Printed in Singapore by Uto-Print

To my wife

Julia Margaret Dobson

Contents

Preface

I have often said how world civilization would not have advanced to where we are today if the Earth had always been a cloud-shrouded planet. The sun, moon and stars have given inspiration through the ages, and we must recognize the long stages of development in the history of science.

We now know that Stonehenge and the Egyptian pyramids were built to mark astronomical cycles, and the sky gave the input to that early science and technology. We marvel at the ancient knowledge shown, and we wonder about the exact technology invented to handle 50-ton blocks of stone.

Each big leap in our understanding has depended on a new communication technology. The Space Age in which we live has come through computers, and satellite systems. In the next stage will there be some form of communication with civilizations beyond the solar system?

Readers of the first edition have appreciated the inclusion of computer programs which enhance my subject more than words can do. Here I am indebted to the late Dr. LeRoy E. Doggett of the U.S. Naval Observatory for mathematical comments. Instead of depending on a floppy disc of unknown content, this book contains algorithms, and spreadsheet examples for hands-on computing.

Finally I must thank my wife, Julia M. Dobson, whose perspective as an expert in the Teaching of English as a Foreign Language, and whose personal knowledge of 23 different countries have added immeasurably to the book.

Gerald S. Hawkins
Washington D.C.
April 20, 2002

Preface to the First Edition

This book is based on lectures given at the Phillips Exeter Academy in New Hampshire, and the Smithsonian Institution and the Cosmos Club in Washington, D.C.

Mindsteps to the Cosmos reach across so many different disciplines that I have not been able to find one person who could act as a mentor in all the aspects. I have long researched the idea through the literature, at conferences and in correspondence, and I have added new material as a result of field trips and expeditions. From 1970 onward, Dr. Labib Habachi of Cairo has been an invaluable guide into the hieroglyphy and monuments of ancient Egypt; Dr. LeRoy E. Doggett of the U.S. Naval Observatory has advised on the computer programs; Dr. Evan Hadingham has generously shared his expertise on megaliths and the European Stone Age; Charles Reed, Esq. has been keenly interested in the calculator programs and the mindsteps; Professor Owen Gingerich of Harvard University has helped with astronomical data and the solar and planetary longitude tables that he coauthored with William D. Stahlman; Dr. Carmen Cook de Leonard of Mexico introduced me to the astronomy of the legend of Quetzalcoatl; Miro Morville has helped with translations; and Pamela J. Dobson, anthropologist, has read the entire manuscript from the viewpoint of her discipline. To these, and others who have helped directly and indirectly, I am grateful. My wife, Julia, has encouraged and followed my lecturing and writing at every stage.

Having spent many years in academia as faculty member and then administrator I fully realize one can't please all the professors all of the time. Nor should book reviewers expect to find that the materials of this book have been exhaustively treated in the technical journals, for much that is presented here is new. For any errors I accept responsibility and offer apologies, but with such a broad theme as cosmic mindsteps one is bound to find a detail where expert A disagrees with expert B. The anthropic principles of Chapter 13 are one example, the precise reading (if one is possible) of the carved stela of Bekhten is another.

Nor do I think the broad sweep of the mindsteps hangs or falls on a hair-fact. I present this theory of the mindsteps as an astronomer's view, and I hope it is accepted with that understanding.

Gerald S. Hawkins
Washington, D.C.
April 20, 1983

ONE
Up from Zero

Humans are unique. There is nothing on earth gifted with intelligence to equal Homo sapiens's. We are born of the cosmos, but is there a place for us in this glittering jewel box of galaxies and stars? The astronomical clock ticks in measured beats, and life cycles pulse like beacons on our delicate planet. Human minds open and reach out, but is there a cosmic destiny, and will it be fulfilled? Only recently have we been able to trace faintly our first steps as Sapiens, the Intelligent, and those vital threads before that time.

Modern Sapiens took over from another type of human being, Homo sapiens neanderthalensis, a type named for the Neander River in Germany where the first specimen was dug up. They appeared before the Ice Age, around 100,000 years ago when it was warmer and the climate produced a food-ladened land. At first the Neanderthals were classed as beetle-browed brutes, but we now know differently. The bumps in the brain case of the skulls show that Neanderthals, like the species before them, had developed human skills, including a spoken language of sorts. They understood the plants and the game animals, the clouds, weather and terrain. Those early humans had an environmental outlook which enabled the Sapiens species to survive. But Neanderthal made mindsteps toward the earth, not the cosmos.

This is the broad picture. Archaeologists dig up the hardware, but they lose the software—the ancient mind slips through their fingers. Ideas, information, understandings—these were carried in the now-empty brain cases, transferred by the spoken word from one brain to another. Communication was vital to survival. A good idea survived and replicated in the memory cores of the brain; a bad one perished. Idea A plus idea B produced idea C. But unless idea A connected with the culture as a whole, it perished.

Each little thought is called a "meme." A group of people share a set of memes. It is like sexual reproduction, where physical characteristics are transmitted by genes, and a cultural gene pool develops. Scandinavians are usually blond and blue-eyed; Mediterranean people tend to be brunette and dark-eyed. In the same way a meme pool develops and breeds in the brain and is preserved in the collective memory. A gene pool interacting with the environment shapes the body. A meme pool interacting with the cosmos shapes the mind.

Today laser beams and cables are fast and direct, but face-to-face communication in the days of the cave was also effective. Today the collective brain is expanded by books, computers and data banks. Billions of memes are stored in hardware chips. But in the days when writing had not yet been invented, the memes had to be carried in the mind. They were passed on in ballads, poems, songs and stories. They were stored in perishable software.

Memes are very important: Religious beliefs have sustained whole populations in times of need and trouble. A faint hope for the future will keep the pulse flowing against all odds. Knowledge of a key fact has pulled victory from the jaws of defeat. On the other hand, a single potent meme could appear, spread, take hold and wreck the world. At least that is what British astronomer Sir Fred Hoyle once said at a lecture in California. His remark caused a stir. Someone in the audience put him on the spot and asked him to give an example. He paused and said, "Fortunately, I can't think of one at this very moment!"

Charles Darwin's survival-of-the-fittest theory applies only to physical characteristics, the genes, but it would be reasonable to add to this the fitness of the thought pattern. Let us hope that thinking hominids have developed memes that increase the chance of survival. Earthlings have made it through almost to the 21st century. So far, so good. But will we in the future stampede our species over the edge of a cliff because of a wrong idea? Let us hope not. Survival hinges on the fitness of the meme as well as of the gene. There will be a tendency for a successful pattern of behavior to be adopted and repeated generation after generation. Conservatism as a meme is more secure than innovation so long as things do not change. Within the sapiens species, Neanderthals hold the record for conservatism. They were creatures of habit.

European Neanderthals scratched the surface of flat stones. A slab from La Ferrassie, France, shows about 18 shallow holes. These are called cup marks. They might have been a crude numbering system. The left hemisphere of the brain deals with numbers, language and logic. It is the quantitative, computerlike, fact-processing side. We know this hemisphere was developed in Neanderthal because of the language bumps in the roof of his skull.

But what of the right hemisphere, the creative side, the nonscientific, artistic, romantic side? This hemisphere is the main source of imagination and creativity. The only clues we have are in the artifacts, the handcrafted Neanderthal possessions. For tens of thousands of years they made the same kind of small, polished, spherical pebbles. If these were not money—and who would need money with the free supply of food at the cave door—these pebbles must have carried a meaning. They were part of an idea passed on through the centuries. At one site another mysterious clue has been dug up. It is a perfect flat disc of white flint, carefully fashioned for a purpose unknown. Other objects are less of a mystery—they were tools and weapons.

Death and its mysteries were recognized. They buried the dead with red ocher especially collected from iron ore deposits, and they provided food and grave goods for the dead. They may have believed in a world of spirits and the existence of a life after death. Perhaps there was a fear of the dead coming back to life and causing them harm, because some departees were buried with legs drawn up and tightly bound. One old Neanderthal male had been laid to rest in a site in present-day Iraq on a bed of spring flowers. At least 8 pollen types were identified, and each flower has now been recognized to have medicinal value. All this new research shows that Neanderthal was not the Stone Age brute previously supposed, but a real human with beliefs, with knowledge of the earth's environment, and with a proven capacity to survive. But Homo sapiens neanderthalensis looked to earth. Those that followed looked to the sky.

Neanderthal was followed by Homo sapiens sapiens, the sub-species to which we belong, called "Sapiens" for short. The first specimen to be dug up was a 50-year-old male at Cro-Magnon in France, and so the new sub-species is often called "Cro-Magnon." Actually there may not be any fundamental difference between Neanderthal and Sapiens. According to a special committee of scholars, a Neanderthal properly dressed, washed and shaved would pass unnoticed in a modern city crowd.

Sapiens's brain 40,000 years ago was about the same size as our brains today and probably functioned similarly. The brain has several layers arranged around the central stem, with the outer layers more complex than the inner. From the point of view of evolution a delicate structure like that must have taken millions of years to develop. It carries the marks of its slow evolution. The R-complex near the center of the brain is inherited from the reptiles. A human reduced by some terrible accident to using the R-complex would have no perception of cause and effect and would live a life of daydreaming and urge satisfaction. The outer layer, the neocortex, is the crowning success of evolution. It is the key to our perception of the world, from the near environment to the far-flung cosmos. It analyzes, sifts information, stores facts and creates things that never before existed. The left side of the neocortex is analytical, the right side is intuitive. Both sides communicate through interconnecting nerve bundles, but essentially there is a separation of functions.

Many of the world's religions hold that man was made in the image of God, meaning of course his mind. It is the sapient mind that separates earthlings from other animals, and this difference is accepted as fundamental and is taken as an act of grace or a result of divine creation. The mind is located somewhere in the complicated and still scarcely understood processes of the brain. If the mind has substance, then it is chemical and electrical. If it is more than this, then it is beyond the realm of science. The left hemisphere with its neurons and nerves functions very much like an electronic computer. "Man was made in the image of God, and the computer was made in the image of Man," or so it is said. But that does not take into account the right hemisphere. It has neurons and nerve connections as does the left, but it tends to be unpredictable, uncodable, quite unlike a computer. I believe the intuitive side was the most important side for the various Sapiens cultures. It gave them a world within a world and enabled them to adapt.

Sapiens men and women worked together in a close-knit life-style, moving to sites where the best food was available, and building huts, cooking places and animal traps. It was a life where demons were real and spirits were in everything. When things happened, they were caused by the great unseen. The women may have outranked the men. Of the numerous figures carved so carefully in bone and ivory, most are female. We do not know why these were carved. Venus figures they are called, with exaggerated breasts and large round hips. Perhaps the shape symbolized human power. Gaston Lachaise's modern sculpture "Triumphant standing woman" captures this optimism and supremacy. Auguste Rodin's sculpture "The thinker" represents the mind with its searching question: "Do we humans have a place in the cosmos?"

Those first thinking earthlings lived in a garden of Eden. It was warm when Neanderthals came on the scene and colder when Sapiens appeared, but the climatic changes were slow and the sparse groups and families could choose to go where the living was best. There were no walls or fences to the garden. They had winter quarters in the warmer south, and summer places for hunting the game. The cave dwellers were warmed in winter and cooled in summer by natural air conditioning. It was not the barren tundra of the North Slope of Alaska. There were trees and pastures, clumps of pine and birch, with grazing for the herds in between. The southern faces of cliffs provided warm glens, and natural springs flowed at the mouths of the caves. The food-population balance was tilted in favor of food. Individuals met with accidents (there are many examples of broken bones) and illnesses (some of which could have been cured by herbal medicines), and occasionally a small group would make a fatal mistake (like the half-dozen who ventured under a cliff overhang during a spring thaw and were crushed). But the species, being more than the sum of the individuals, was well adapted to survive, and for millennia the planet was an ideal environment for the slow nurturing and establishment of earthlings.

If Neanderthal was a creature of habit, Sapiens was not. Sapiens improved the hunting methods, performed the first crude surgery, used fire to cook foods and remove the toxic alkaloids from vegetables, listened to the music of the first (bone) band, and invented art.

Stone Age music ensembles found on the settlement floors conjure up the idea of dancing and entertainment. There was plenty of leisure time and the days and evenings were long. But it is more likely that the music was for a purpose. A set of pipes and drums nearly 20,000 years old was found in the U.S.S.R. Drum beats are a way of tapping out the passing of time and of memorizing numbers. Time itself might have been a god concept. At the social level rhythm has the practical effect of giving a feeling of security.

I do not wish to get involved in the evolution argument: Did it take place slowly over the aeons, or did it move forward in a spectacular jump—could a hopeful monster appear on the scene by mutation and find a niche? Was there a sudden change in the brain caused by a quirk, or was the modern mind a slow and steady 3 million years in the gestation?

Language capability was built into the system before the time of the Neanderthal, judging by the hollows in the fossil skulls. Nowadays a baby seems to have a precoded recognition of language functions, of the difference between words for things and words for actions. It is just possible that with modern Sapiens evolution suddenly produced a brain with precoded ideas, ready to be brought into play. Nobel laureate Gerald M. Edelman once speculated in a lecture at Rockefeller University that brains at birth came already filled with ideas. The only problem is to match these ideas with the world outside. The human body is coded at birth to fight existing viruses—even new bugs yet to be produced by transmutations in nature. This theory came from his prize-winning work on immunology. If it were not so, he said, our species would have died out from plague long ago. In the same way, he argued, we have memes embedded in our subconscious ready to deal with the reality yet to come. Perhaps so . . . like the unfolding of a flower to reach the sunlight.

Canadian psychologist John M. Kennedy believes that modern Sapiens is born with a precoded drawing ability. He found that remote tribes that had never made drawings could easily recognize 3-dimensional objects in 2-dimensional drawings. Test subjects who were blind from birth could identify an object in a braille pattern, even though they had never seen, but only felt, the original. If Edelman and Kennedy are right, then people who share the same culture might respond suddenly and naturally to an idea that comes along at the right time.

The Neanderthal culture did not have any pictorial artists. No drawings have been found. Was the art-meme lacking, or just dormant? Pictorial art erupted with Sapiens. Whether it was a stimulus from the outside or a deep urge welling up from the right brain hemisphere we do not know, but this art first appeared about 35,000 years ago. The Lascaux caves in France and the Altamira in Spain are examples. Those first artists mixed water, fat and mineral pigments, reached the peak of the ceiling (perhaps with high scaffolding), and crawled into dark recesses with flickering torches to leave for the world a gallery of long-gone animals, spirits and mysterious symbols.

As with all good art the pictures have something within them that reaches out to join the artist with the viewer. Yet thousands of years later we are in a different world, pressed down with crises and a whole overlay of ideas from the past and present to cope with. There is a vast cognitive gap, and we are not equipped fully to understand that sudden outburst of art in the caves. Some prehistorians have called it simply "art for art's sake," meaning that it pleased the artist and the viewer, but nothing more. Others see in the drawings evidence of sexual symbols, and still others believe the drawings were hunting magic, voodoo rituals or representations of a world of Stone Age demons. Perhaps there is a little truth in each theory, but one thing is clear: those cave dwellers had made a discovery, an invention—pictorial communication. The Neanderthals could exchange ideas by talking, but Sapiens could communicate by pictures. The technological step was dramatic. It was like going from radio to television, from phonograph to videotape. Neanderthal language disappeared into the air, but the Sapiens message would last as long as the pigments stuck to the walls of the cave.

Cave drawings, possibly of the sun, La Pileta, Spain.

In some caves and rock shelters there are circles, some with rays. These might be sun symbols carefully drawn. Others might be stars. If this is so, the creative Stone Age artists, depictors of earth and animals, were also watching the heavens. Their boundary was not the local environment of the Neanderthals, but the sky above. They went from things within their grasp to something beyond their reach. Perhaps, and if so it was an important step. They had a sense of wonder about the cosmos.

As if to signal that the left hemisphere was alive and well, the cave artist added geometrical patterns, snake-like rows of colored dots, and numbers. At least, that is the only way I can describe the "combs" and "rakes"—they seem to be a record of numbers. My first visit to one of these caves was in the farm of La Pileta in southern Spain. The lines on the walls were as fresh as when they were made, shut off and preserved from the outside world for thousands of years. The guide held up a kerosene lamp to help me see the horse and other figures. To my eyes the combs and rakes were recorded numbers. They were very popular markings, more frequent than the animal drawings. On one wall I saw all the numbers from 3 to 12 as if the artist were counting. The signs could of course be decoration or nothing more than art for art's sake, but my vote goes for the quantitative interpretation, a necessary development in the mind for any understanding of the space and time which underlies the cosmos.

Cave art is the earliest evidence we have for human cosmic awareness. The act of looking, thinking, and communicating visually is a prerequisite for any connection between humans and the astronomical universe. It is a receptive state, a beginning, an awakening, a cosmic mindstep zero. Without reaching this stage of development there can be no progress to relating humans to the greater cosmos. No answers to the question can be found unless the question is asked.

But why mindstep "zero," why not call it "one"? This is because of a nicety of mathematics. Zero is more than just a word for "nothing." It is a number in its own right, equally important with 1, 2, 3, etc. Actually "0" is more important than the other numbers because it stands in the middle, dividing all the negative numbers from the positive. For example, take the steps in a staircase. The ground is 0 and the first step is number 1.

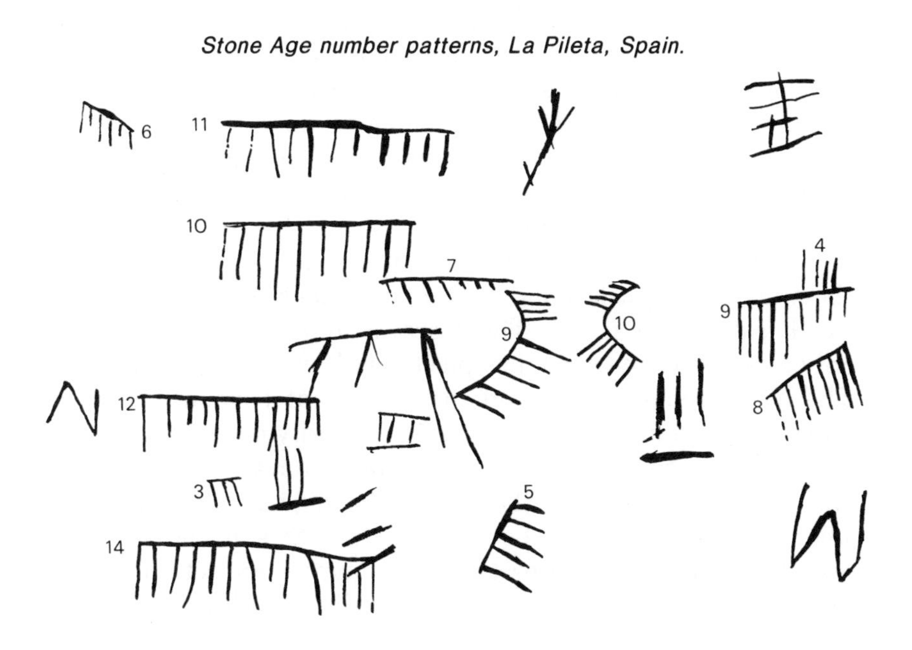

Stone Age number patterns, La Pileta, Spain.

Going below, the first step would be "minus one," the next, "minus two," etc. Mathematically there is no way to leave out the "0"; it is a threshold, an origin. When the Ice Age people looked up to the sky and first began to notice things astronomical, that was their starting point, and zero is the correct label. Before that the steps were negative—of the earth and not of the cosmos. Mindstep 0 was the Threshold of Cosmic Awareness.

We would have a lesser opinion of those Ice Age people if they had let the ocher and soot lie idle. They published in a grandiose way. Their paintings are delightful as well as informative. They also published by artifact. They left behind hundreds of small carved and incised bones. Some of these are classified as art, but others are different. There can be no doubt about the age of these objects; the bones are found in the deep layers of the caves, and the lines were cut into bone fresh from the kill, before it began to petrify. Some of the bone and ivory came from mammoths—those woolly elephants that became extinct some 10,000 years ago.

Most of the paintings are connected with the environment, and the art speaks of animals and people. But the carved bones are not so easy to comprehend. If it is art, then it is mathematical art. If it connects with the environment, then the link is difficult to see. These fine incisions can hardly be hunting tallies because it would seem to be out of proportion to record the killing of a mammoth with a single scratch no longer than a dime. I think these sequences have something to do with time, like the unfolding of a folktale. The marks could be cues for stories or ballads in the form of memory sticks. But if they were, the words have long ago vanished into the air.

I can vaguely see a moon story in the Gontzi bone. This is a short mammoth tusk found at Gontzi in the Soviet Ukraine. One can hardly say it is art. There are no representational figures, only a series of fine scratches placed with watchmaker precision. The bone probably marks off the passage of 4 moon-months, or lunations. The month which I have arbitrarily labeled "month 2" is the easiest to see. There are 28 marks for the visible phases and 2 short scratches at the end of the sequence for the nights when the moon was in the direction of the sun and therefore invisible. Full moon and first and last quarter are set off with an extra long line. There are 3 other months recorded on the tusk.

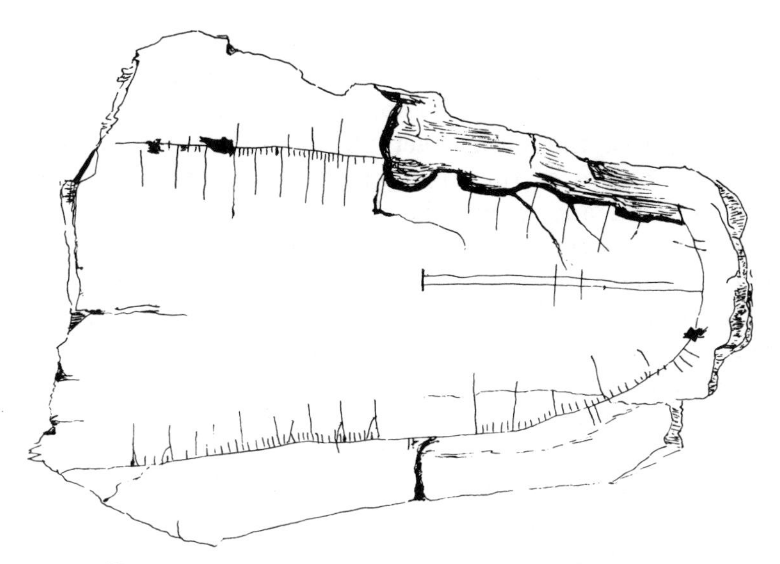

Fine markings on a mammoth tusk, Gontzi, U.S.S.R.

The 1st and 4th months are divided up, but differently, not in groups of 7. Unfortunately the bone is damaged where the 3rd month should be, but enough shows to convince me it was there.

Could the Gontzi bone be a sort of astronomical record? Does it mark out the changes in the moon so far back in time? We must remember that it is anywhere from 10,000 to 15,000 years old. So far only one Gontzi bone has been discovered. Other carvings have been identified that might fit into a moon pattern, but none compare with this 4-month sequence. It stands alone. One swallow does not a summer make. But there is no other explanation for it.

In the rock shelter of Abris de las Viñas in Spain a shaman or cave dweller or hunter painted a picture that has lasted for nearly 10,000 years. It seems to be a god-like form surrounded by a pattern of spots and crescents. It may be more than art. It may be symbolic. As an astronomer I see the possibility of a thinking beyond mindstep zero, beyond the simple awareness of the previous millennia. The spots painted around the figure are in groups of 7 and look very much as though they were meant to count off the 7 nights of the quarter of the moon. The 3 separate spots (or 4, depending on how the artist has designed the rows) are the nights when the moon is at the full phase and at its brightest. Although the almanacs nowadays print the day and exact time of the instant of full moon (the split second when its center is precisely opposite the sun), it actually doesn't look much different the night before or the night after. To the naked eye it looks equally round and bright. For the ordinary observer there are really 3 nights when the moon looks full. These nights of "fullness" are the 3 spots at the top of the pattern of the drawing.

It is cave art. It is attractive and eye-catching. An abstract creation. But to the astronomer, it speaks astronomy. The pattern of dots is a representational tally for counting off the days of the moon-month. Either it is an almanac-cum-counting device, or a pictogram to enshrine a known fact of the cosmos. In either case, to me it is one of the first pieces of evidence of earthlings explaining to themselves the phases of the moon, and the date is fixed at about 7,000 B.C.

There is another moon-month painting in a nearby cave shelter at Canchal de Mahoma. It dates back to the same period and has 27 spots which change from a crescent to a round shape and back to the crescent. (Here I count the "v" as two marks joined together.) The faint straight lines at the lower right also have an astronomical interpretation. These are the evening and morning when the moon is too close to the glare of the sun to be visible. Like the first example, the artist has drawn something, maybe a figure, at the center. This could represent a god-figure, or even the concept of the earth as the center of the universe with the moon traveling around it.

We would like to know if the two moon drawings are related. Is there a cultural connection? They are both in Spain, one at Canchal de Mahoma and the other at Abris de las Viñas. The locations look close together on a small-scale map, and time-wise the drawings belong to the same archaeological layer—Azilian. But distances on foot over rough ground are different from distances on paper, and the Azilian culture lasted for a thousand years or longer. It is unlikely therefore that they were done by the same hand. This is borne out by the style. The Canchal de Mahoma work is more primitive in execution than the Abris de las Viñas. The Canchal one is mixing visual experience (the way the crescent looks at sunset) with the idea of the passage of time (the sequence of dots). There are 27 marks, one for each of the nights when the moon could actually be seen in the sky. When

The earliest identified notations of the moon-month: (Left) Canchal de Mahoma. (Right) Abris de las Viñas, Spain.

the moon was gone from the sky the artist was confused. There are two rather feeble marks for the moonless nights, but these are not placed in the proper sequence between the crescents. The thin lines are placed hesitantly outside the "orbit" which circles the god-figure.

In contrast the Abris de las Viñas drawing is firm and confident. It says: "There are 30 different shapes of the moon; on some nights it is invisible, but it is there nevertheless . . . dark moon, dark moon, waiting for the light."

The two moon pictures, if that is what they are, tend to confirm each other. Not that two swallows make a summer, but neither picture can be called a throw-out piece of art. Is it the first sign of human awareness of the orbital motion of the moon, a sense of wonder about the changes of the moon's shape? If so, it was a big step, because time and space were also involved.

If the two drawings were done by the same hand, the idea could have come from the mind of one person, unshared by others in the tribe. If so, it was a maverick meme that perished with the artist. But it is more likely that we are looking at part of a larger idea shared by the peoples of the Azilian culture.

Not much is known about the Azilian mind. They did not

paint bison and reindeer and are placed somewhat lower on the artistic scale than the cave artists of Altamira. Was it that the brain's creative hemisphere was now resting?

Instead of fine art, those Azilians left behind the "pebble puzzle." They decorated pebbles with dots, lines and zigzags—thousands of them. This peculiar habit went on for generations. Whether the Azilians were recording something numerical or were crudely analyzing a problem we do not know. The decorated pebbles, like the moon-dots, might well stand for a sequence of ideas which everybody who was an Azilian understood. Nobody understands the markings today. Perhaps mythologists or numerologists will one day decode the meaning. It was as though in the long course of prehistory the left hemisphere of the brain had switched on.

Was the orb of the moon a god-figure to the Azilians? Is the blob at the center of the moon-dot patterns representative of the earth, or gravity, or some controlling influence on the moon's path in the sky? Do the rows of dots on the pebbles and their breaks and turns represent incidents in some long-forgotten Azilian folktale?

Swiss psychologist Carl Gustav Jung said we are all born with archaic horizons built into our minds. Layer by layer we can go deeper and deeper into the subconscious where the prehistoric memory survives. There is one horizon equipped with all the emotions of the hunt, another has a coded program for worship of the sun, still others connect us with the world of spirits and demons that our remote ancestors took for real.

Jung's "horizons of the subconscious" are conceptual, not physical. The layers are not necessarily located in any fixed part of the brain. But they are shared, embedded in the collective subconscious of a civilization. Carl Sagan's *Cosmos* showed the sense of awe and wonder that Homo "modernis" has. This cosmic zero, mindstep zero, probably began in the caves. If our species had not reached upward from planet number 3 as it has done, we would still be back with the mind-set of Homo sapiens neanderthalensis.

TWO
Mindstep 1

We crawl, we walk, we run. Between birth and age 6 months the brain doubles in capacity, and humans are then equipped for coming to terms with the universe. But this reaching out is a gradual process. I can remember an exploration adventure at the age of three. I walked by the beach and the sea along Great Yarmouth's promenade on the east coast of England. I had slipped out of the yard full of comfortable toys and wandered unnoticed for about a quarter of a mile. Suddenly I was in an enormous world of flat red paving which stretched away from me in all directions. Like a comet at the end of its orbit I had reached my infinity. It was a scary experience, but someone managed to find out where I had started from and took me back to the yard. Years later I went back to look at that spot. The sidewalk was unchanged in front of the Wellington Pier. That red infinity of a 3-year-old now looked quite ordinary.

Aged about eight, I noticed the stars. Until then I looked no farther than the clouds. They were unreachable, far away, and they moved. I knew that thunder rolled around up there and lightning bolts came down. Again it was at the Wellington Pier. The summer beach-show was over in the Pavilion and the theater crowd stood to watch a fireworks display. Faces were turned upward with open mouths like crocuses in the spring fields, and I could see a small patch of sky between the heads. When the rockets faded and the blue-brown smoke drifted away, I saw the stars. Millions of them, it seemed, more impressive than the fireworks, and they were hung in the roof of a great velvet dome.

The starry sky is the Biblical "vault of the heavens," or the old "firmament" of the King James version. It is a smooth surface which makes a comfortable, secure cover for people below. Earth and sky are separate. When a whole civilization recognizes this, it is the first step—mindstep 1. I doubt that any culture classified as a civilization has ever existed without it.

This, of course, is from the astronomical perspective. There are other steps of the intellect not related to the sky—the taming of fire, the adoption of legal codes, the invention of the wheel—but I am speaking of astronomy, the broad advances that seem to be fundamental in our search for a meaning to this tiny earthling existence within the framework of the universe. A mindstep is a massive change of thinking that alters the relationship of humans to the cosmos.

It is difficult to fix the dates because the hard evidence is so sparse, but the beginning awareness of the heavens must have been followed by an attempt to explain what was seen. The initial stage, the recognition of things in the sky, took place long ago in the Ice Age. I equate that awareness threshold, mindstep 0, with the appearance of art—symbolic writing, the invention of the picture. The date would be about 35,000 B.C. It is just possible for mindstep 0 to be further back in time, say with the Neanderthals around 100,000 B.C., but the marks and scratches on the stones show no evidence of their looking at the sky.

In going from "awareness" to "explanation," ancient cultures were moving forward by one mindstep. Mindstep 0 began the Age of Chaos as earth and heaven intermingled, and there was no way of accounting for what was observed. Then the earth and the heavens were separated, the realm of the gods was established in the vault overhead, and the cosmos was explained by myths and legends. Cronus with his sword separated mother earth (Ge) from father of the heavens (Uranus), and then mindstep 1, the Age of Myth and Legend, began.

Mindstep 1 is linked with the development of myths and legends as explanations for the behavior of the gods of the sky. I tentatively place it at 3000 B.C., at the time of the invention of writing, though it could be earlier, perhaps as early as the paintings of the moon phases in the rock shelters in Spain about 7000

B.C. One can hypothesize that the pattern is a moon calendar, and that a legend was told about the changing shape of the moon, but it is only a hypothesis, and the first hard evidence of astronomical stories comes along only with the appearance of writing. With that invention the stories could be preserved and transmitted more faithfully. Before the invention of writing, the sky-god legends would have had to be transmitted by word of mouth and held in the memory. It seems impossible now for researchers to trace back so far in time on the basis of oral tradition.

Mindstep 1 took place at least 5,000 years ago, yet there is still a need to have something explained by a story—particularly for children. In the 1960s I was concerned about city lights and TV and losing a link with the past, so I wrote a children's book about the phases of the moon called *The Moon Tonight.* George Solonovich illustrated it beautifully and it contained a little poem which began: "The moon tonight is new, it smiles a golden smile," and it ended: "The moon tonight is dark, it sleeps behind the night, dark moon, dark moon, waiting for the light." It was designed to introduce the idea of the moon and the way it moved and changed its shape. But I was surprised to find that the idea was already subconsciously familiar. One couple had adopted an American Indian girl at birth. When they read her the book at bedtime and showed the pictures (she was 4 years old at the time), she said: "I know all that." Another boy across the street from me read the book and looked at the moon. He saw it change its shape each night and he liked the way it followed him as he walked around. (This is an effect called "parallax." Because the moon is so far away it seems to move over the tops of the trees as the observer moves.) One evening it was not visible, being at last quarter and rising after midnight. He was very upset. He had lost it. "Where's my moon?" he cried.

The moon by itself, even though it is the brightest night-sky object, is not all of the cosmos, but it is a very important component. In some cultures it was singled out to be more important than the sun. As the ancient Peruvian Indians argued, the moon can change its shape and the sun cannot, and it has a face. The moon can be thought of as a white light to chase away the evils of darkness, a changing, moving god, a calendar counting device, a thing to be possessed, or more. But the moon is only one of the

Full moon and homing ibis.

many objects up there to be recognized, and if the cave people were watching the moon they were taking in only one small part of the cosmic story.

Giorgio de Santillana, a humanities professor at the Massachusetts Institute of Technology, believed that cosmic understanding goes back tens of thousands of years. He said the star patterns were picked out, the zodiac recognized, and the shift of the celestial pole from north star back to north star had been observed and noted in the Ice Age. Now, this motion of the pole takes about 26,000 years for one complete cycle. I find it difficult to go along with de Santillana's suggestion when I look at the meager notches in the bones and the very few pictograms of the sun and moon in the caves, though of course it is always possible for such knowledge to have been passed on by oral tradition. The time span of our forebearers was certainly long enough, but the solid evidence is not there.

No, I think mindstep 1 was made between 5000 and 3000 B.C., long after the ice had retreated and after agriculture had replaced hunting as a way of providing food. There is an astronomical clue in the signs of the zodiac.

Because the earth goes around the sun in a nearly circular orbit, the sun gives the illusion of moving day by day across the starry backdrop. Earthlings are an audience in a vast theater and they watch the movements of the actors on the stage. The disc of the sun moves from right to left and it takes 1 year to complete the journey. Of course, the starry background cannot be seen behind the sun (excluding those rare occasions of a total solar eclipse) because the glare of sunlight blots them out. But somehow or other the ancient stargazers figured out the track of the sun despite this difficulty. The track was marked out by special star patterns, the constellations of the zodiac. We suppose early stargazers looked carefully at the stars in the western sky at sunset and in the eastern sky at dawn. In this way they knew the sun's disc must be somewhere in between, and they could figure out the position. The moon also travels around the zodiac, taking 1 month for the journey, and this gives a second method for finding the constellations of the royal road of the zodiac.

In astronomy there is always a first and simple explanation, and then extra facts have to be added to give a more complete

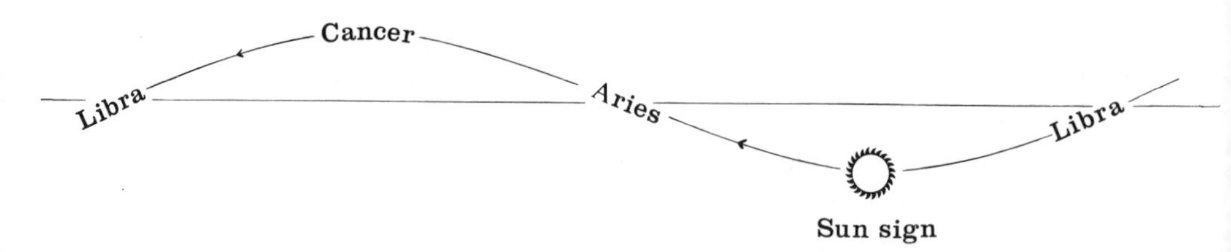

Yearly path of the sun through the zodiac, along the ecliptic.

picture. This is because the universe itself is not simple. It contains complications and surprises.

One complication is the spin of the earth on its axis, but this caused no difficulty to the ancient world. So far as the earthling audience was concerned, the seats were solidly fixed and the stage resolved around the viewer once per day. The whole performance—star backdrop and actors—moved daily from left to right. The sun, moon and stars rose in the east, traveled across the sky and set in the west. There were many "explanations" for this rotation, but it was not a vital part of the zodiac story.

The complication in the zodiac story is the tilt in the axis of the earth. If it were not for this tilt, the zodiac band would follow the celestial equator, directly over the earth's equator. The sun would climb to the same height in the sky every noon, summer and winter. Because of this first complication the zodiac band is inclined by 23½ degrees to the celestial equator, and so the sun climbs high in its tracks in summer and dips down low in winter. The difference between the high point and the low point is 47 degrees (23½ up and 23½ down). Actually we now know from calculating backward that the tilt angle in antiquity was slightly greater, 24 degrees.

The early stargazers could see all this in the sky without any need to know about the earth's axis or orbit. The path of the sun was a given fact of the environment. The track was tilted and the sun took a year to go around. When it was high, it was summer; when it was low, it was winter. Halfway between it was either spring or fall. The belt of the zodiac spun around like a wobbly wheel. De Santillana says the ancient world was very disturbed by this complication. The legends are attempts to explain it. In the beginning the zodiac was designed by the god Saturn, who arranged everything to be aligned correctly, but, owing to some

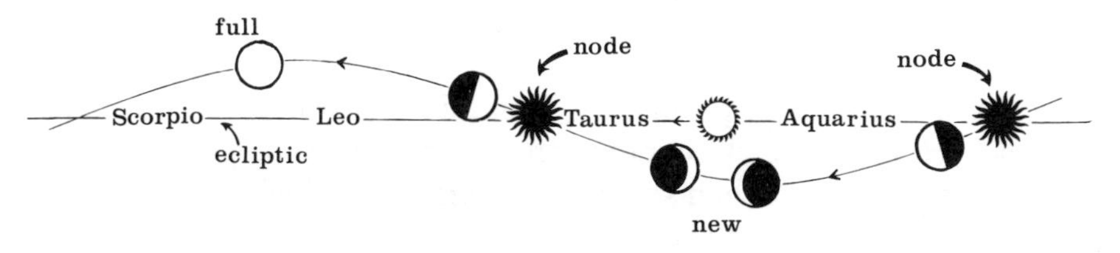

Monthly path of the moon through the zodiac, crossing the ecliptic at the nodes.

great calamity, the zodiac became tilted. The great disorder in the cosmos had appeared.

The second disorder would have gone unnoticed at the time it was designed. That disorder is precession. It gives us a remarkable clue as to the date of the laying down of the zodiac.

The earth's axis holds to the same tilt angle, but while so doing it makes a slow, conical sweeping motion, something like a spinning top when it is slowing down. Called precession, 26,000 years are required for the axis to make a complete circle and return to the starting point. Because of precession the stars of the zodiac appear to skip to the left at a steady rate of about 15 degrees in a thousand years, or 1.38 degrees per century. Technically it is known as "precession of the equinoxes." The vernal equinox is the position of the sun against the background stars on the first day of spring, and the autumnal equinox marks fall. The technical term has these points moving against the starry backdrop, but it is easier to imagine the star patterns themselves to be moving. Precession does not affect the rising or setting of the sun on the horizon, nor does it affect the angle of the sun's rays season by season. But it does change the place of the stars year by year, and they slip relentlessly to the left. If the pattern of the stars had been chosen to make sense or "tell a story" in the year 5,000 B.C., it would be out of sync by the year 3000 and still worse in our own time.

This is exactly what has happened. Precession has shifted the entire zodiac. We use the old constellations, handed down to us by the Greeks, but they do not fit the seasons today. Like King Canute who at the Dee estuary proclaimed that the tide would not come in, the ancient signs of the zodiac were proclaimed to be fixed. But the tide did come in over Canute's feet, and the stars did move nevertheless. The constellation of Aries is where Gemini used to be 5,000 years ago, Leo has replaced Libra, and so forth.

At what date in prehistory, then, do the star patterns make sense? Here we need cooperation between mythologist and astronomer. To me it seems to be the period just before the invention of writing, before the emergence of the great civilizations. At that epoch the constellations were in tune with the seasons of the year. The interpretation goes as follows:

In 3000 B.C. the sun began the yearly journey up into the northern sky when it passed through the constellation of Taurus, the bull. The horns are directed to the left and the disc is prodded (tossed?) on its upward path. This was at the vernal equinox, the month of March in our modern calendar. In June the sun reached its highest point, marked in 3000 B.C. by the constellation of Leo, the lion. The mythical king of the beasts, the animal of the hot lands, Leo was rightfully placed to receive the sun at the highest peak of the zodiac. When the sun passed through the autumnal equinox it passed through Scorpio. Old charts show the scorpion with claws poised facing to the right, ready to clutch the disc and drag the sun down into the southern regions. Three months later the sun has reached the lowest point, the winter solstice, and here it is projected against the stars of Aquarius, the water carrier. Aquarius is always depicted pouring out water on the earth—the rains of winter.

I believe the zodiac signs were chosen deliberately to fit in with the seasons, to tell a story, to give an explanation. One cannot argue the other way around—that the star patterns are so obvious as to suggest their own story. The images are not obvious at all. The stars do not mark out a lion or water carrier. The grouping of stars in each zodiac sign is a "forced fit" to the image required. Most people see Leo as a sickle with the bright star regulus at the handle. But the designers did not want a sickle at the high point in the zodiac; it had to be a lion. Aquarius is a small triangle with a fourth star at the center, more like a pyramid than a god with a water jug. Gemini is a rectangle, not two brothers. Aries and Pisces are nothing like a ram and fishes. The bull comes closest to being a true image in the sky. The head is marked by the V-shaped Hyades star cluster, and two stars show the tips of the horns. A bright red star, Aldebaran, is the (evil?) left eye as seen from earth. But that is all. Horns, but no body, no tail or important parts.

I have begun the story with the bull in the spring, but there may have been a different month for "New Year's Day" depending on the culture. The starting place is not important because the drama of the sun's journey remains the same. The signs of the zodiac around 3000 B.C. foretell the seasonal events wherever one starts. It is an eternal cycle.

The amazing fact about the zodiac is that it is nonattributable. We don't know the inventor, be it man or woman, priest or king. The work is stunningly noteworthy, but there is no name to call, no one to step forward. By the time the 12 signs became part of the written record, the zodiac was already old. The Egyptians carved the zodiac on the stone of Dendera around 30 B.C.; it was well known to Hipparchus on the island of Rhodes when he made the first star chart around 100 B.C.; the Chinese mapped it in the Chou period around 400 B.C.; Homer in his epic poems alludes to it around 700 B.C.; and there are references in the earliest clay tablets of Mesopotamia. Probably the zodiac doesn't have a single author. More likely the concept was put together over a long period of time, over many generations, with adjustments and improvements made bit by bit. The final construct was passed on by word of mouth and became part of the collective knowledge. As well as being a practical calendric device it gave a mystical understanding of the movement of the solar disc and the passage of the seasons. The signs were also coupled with astrology and the occult.

The names of the signs of the zodiac as we use them come from the ancient Greeks. Zodiac is a Greek word meaning "circle of life," or "little animals." All the constellations are animal or insect except Libra, but even Libra can be taken as the balance for weighing the soul. We use the Greek names because of the heritage of the classics, but the zodiac concept was probably developed in the Tigris and Euphrates valleys, in the Sumerian and Akkadian civilizations before the time of Babylon.

Ancient Mesopotamia is now part of Iraq, Iran and Jordan, and the old civilization has vanished. The cities are buried under the rubble of scrub-covered hillocks; the ancient irrigation system is clogged with dry silt. But 5,000 years ago there was a flourishing agriculture along with sheep and cattle raising; there were palaces and temples, and powerful rulers and laws.

The region was the birthplace of writing. At first, about 8000 B.C., clay and stone tokens were used as records of ownership. A sphere was a token for a bushel of barley, a cone stood for one-sixth of a bushel. By 3500 B.C. the inhabitants had begun to store the tokens inside hollow clay balls for safekeeping. Next they marked the clay balls while they were wet to show what was contained inside, and in so doing they learned how to mark clay with symbols. The earliest date for crude symbolic writing is about 3100 B.C., when small tablets were pressed with the spheres and cones before the clay hardened. A few centuries went by (which is not a long time for such a great leap forward) and a system was invented for writing words and numbers with a wedge-shaped stylus. This is called cuneiform. In nearby Egypt the hieroglyphic system of writing developed simultaneously. It is one of the puzzles of archaeology as to who copied whom. The cuneiform and hieroglyphic systems are entirely different, and the languages are not the same. The Mesopotamians wrote on clay tablets; the Egyptians carved their several hundred hieroglyphic symbols into stone. Both systems sprang up essentially complete at the outset. The Egyptian system was used almost unchanged for thousands of years. Perhaps each invention was independent, like the wheel being invented twice.

With these advances the focus of intellectual activity had shifted from France and Spain to the Middle East. The caves had spawned art, and the Middle East became the birthplace of agriculture, writing and the rending apart of earth and sky. The same discoveries were to be made in the New World, but a thousand years later. The developments in Chinese astronomy were also later. In fact, China was relatively unpopulated in 3000 B.C.; the sudden appearance of dwellings and villages did not come until around 2000 B.C. in what archaeologists call the "Neolithic Hiatus." Probably this new population had migrated across from the West; there is an East-West connection in the artifacts. Around 3000 B.C., the somewhat isolated population of England was building megalithic tombs and the Stonehenge archways to point to the sun and moon at the seasonal turning points on the horizon. Perhaps this was a showing of mindstep 1. There certainly was a wonder and awareness, but without written evidence (the Stonehengers were nonliterate) it is difficult to equate them in their knowledge to the people of the Tigris and Euphrates valleys.

George Michanowsky dates the development of Mesopotamian culture to the time of a supernova explosion in the constellation of Vela. He has translated a Babylonian cuneiform tablet (around 1000 B.C.) which he thinks is a copy of a Sumerian tablet going back a thousand years earlier. The tablet refers to something special with regard to the constellation Vela and reads: "The gigantic star of the god Ea in the constellation of Vela of the god Ea." Now a supernova is a stellar explosion in which most of the material is ejected and a dense pulsar is left behind. By measuring the present-day expansion of the expanding cloud astronomers have dated the explosion to about 10,000 B.C. Because of uncertainties it could be as late as 4000 B.C. Michanowsky suggests that this super-bright star was seen, was noted for at least a thousand years in the preliterate memory, initiated an awe of the sky in the south where Vela lies and was recorded in the texts of the tablets when writing was invented. This may be so. Certainly an event like that could have been a dramatic stimulus if the audience was ready and perceptive. But Vela is a modern constellation, broken off from the old star pattern of Argo Navis sometime after the time of Edmund Halley (for whom the comet is named). It is a pattern marking the sails of the ship of the Argonauts, and was designed by 18th-century astronomers drawing on Greek mythology. If "Vela" is the correct translation of the cuneiform text, it is difficult for me to see how the ancient word connects with the recently named constellation. I am not against speculation in scientific research. It is necessary. Speculation is the antenna of discovery. Without it no progress can be made, but the evidence based on one supernova is too slim for a credible theory, even though it might be just possible. The zodiac, on the other hand, is a complete picture. It is a clear theme, evidence that we can trace through myths in different languages and in many geographic areas.

Working with the clay tablets is tantalizing. Many are fragmented and many crumbled away to dust in museums before scholars could work on them. Nor are the languages of the tablets spoken anymore. The words have to be figured out from comparisons with later languages. Their authors also used a difficult numbering system—sextuagesimal, counting in groups of 60 for everything, like the seconds and minutes of the clock.

The Sumerians called the stars "the heavenly flock." The controller of the heavens was Sibzianna, "the star of the shepherd of the heavenly herds." Earthly things like cattle rearing and agriculture were important to them and they transferred these earthly activities to the sky. The Babylonians, who came after them, picked up the knowledge, and in turn it was embellished by the Phoenicians, Greeks and Romans. It was a long meme-chain. Our civilization is the end link in that chain.

Today we call Sibzianna "Arcturus." This is Latin for the Greek name "Arktouros," which breaks down to Arktos (a bear) and Ouros (a guard). It is a star in the constellation Boötes, the herdsman. Before passing it down to the Romans the Greeks had switched the role of Arcturus from herdsman of the stars to herdsman of Ursa Major, the Great Bear (the Big Dipper to us). They did this to make it fit in with their legend of Callisto, the beautiful girl who was changed by the gods to a big, ugly bear and lifted to the starry skies. Arcturus is the brightest star in the northern hemisphere; you can find it any evening by extending the curved handle of the Big Dipper. In the present century the light from Arcturus was used to trigger a photocell and open an International Exposition in Chicago, and was used as a navigational star for controlling interplanetary space probes.

The first mindstep up from zero was a big one and it took thousands of years for its impact to become fully absorbed. We cannot assign an actual time or place to this mindstep. It was close to or before the invention of writing, and it may have occurred at several places at different times. But without this mindstep modern Sapiens would have been doomed to an earthy intellect. There was a pantheon of gods overhead. People had awakened to the cosmos "up there." Over the years the idea was played out in myth, legend, religion and philosophy. Judaism embraced it in the book of Genesis; Christianity symbolized it in the vertical bar of the cross as a connecting link between earthlings below and God above; and the holy Koran of Islam is believed to have descended from the seventh and highest heaven to the earth, near Mecca.

We know that this mindstep was taken—our modern understanding is built upon it. There was from that time on a distinct place for mortals in the universe. Knowledge in its wider sense had penetrated into the garden of Eden. Sometime between the ending of the Ice Age and the beginnings of agriculture, earth and sky became separated. Ice shields pulled back to expose the fertile earth, and fog cleared away to reveal the cosmos overhead. Life-force, numbers and music were taken from the celestial sphere. Folklore, myths and legends sprang up as the mind expanded one notch and a rich literature developed. Whole cultures linked their future to the comings and goings of the sun, moon and planets.

Mindstep 1 is lost now in prehistory, but its effects were cataclysmic, reaching and influencing all parts of the world. The step shows most clearly in the civilizations of the Tigris and Euphrates valleys.

THREE
The First Story

John Punnet Peters galloped on an Arab stallion up to the mound of Niffar in central Iraq. He knew it was the site of the ancient city of Nippur. Thirty miles to the east was the mound of Babylon, 20 miles to the west was Shuruppak and 50 miles south, on the banks of the Euphrates, was the ancient city of Uruk, the Erech of the Bible. Peters headed an expedition from the University of Pennsylvania. The task was to dig for tablets. The date, late summer 1888.

It was the "heroic" age of archaeology, when people dug for treasure more than for the cultural mind. The locals also knew the market value of the mound, and before the year was out they captured the "dig" and terrorized the archaeologists. Peters ran for his life, though not before shipping out 40,000 tablet fragments.

The loot (Peters had struck the Nippur library) was transferred to museums in Philadelphia and Istanbul. Scholars cracked the tablet code, aided by an inscription on a rock in Iran written in Old Persian and repeated in cuneiform. What they found was the epic poem of Gilgamesh, king of Uruk. The same long poem has also shown up on fragments from surrounding

Note: The new findings in this chapter are based on the author's lectures in the Smithsonian Institution Resident Associate Program, 1981–82.

sites as far north as Turkey and as far east as Syria and Israel. The poem covers 12 full tablets, each with 6 columns of 300 lines. It is book-length, comparable to Homer's *Iliad,* but was put down in writing 2,000 years before Homer and was recited centuries before that. One tablet contains the original story of the Flood, though the builder of the Ark is Utnapishtim, not Noah, and he lived in the forsaken city of Shuruppak.

Gilgamesh was a real person. He is listed as the 5th king of Uruk and he reigned about the year 2700 B.C., but as a hero-king his name had become attached to earlier mythical figures and events. The meter of the poem and the forced repetition of phrases are the hallmarks of a ballad learned by rote and passed on by word of mouth. The epic of Gilgamesh and his friend Enkidu comes to us from somewhere across time. It is the first story ever to be written. If there is an older story, it has yet to be found. Words floating in the air were caught and baked into clay. Little did Peters know his spade had struck a cosmic story, "The Legend of Gilgamesh," a page from mindstep 1.

A tradesman's seal, about 2700 B.C., showing Gilgamesh and wild beasts.

★ ☆ ★

"Gilgamesh went abroad in the world, but he met with none who could withstand his arms till he came to Uruk. But the men of Uruk muttered in their houses, 'Gilgamesh sounds the tocsin for his amusement, his arrogance has no bounds by day or night.'

"The gods heard their lament, the gods of heaven cried to the Lord of Uruk, to Anu the god of Uruk.

"When Anu had heard their lamentation the gods cried to Aruru, the goddess of creation, 'You made him, O Aruru, now create his equal; let it be as like him as his own reflection, his second self, stormy heart for stormy heart. Let them contend together and leave Uruk in quiet.'

"So the goddess conceived an image in her mind, and it was of the stuff of Anu of the firmament. She dipped her hands in water and pinched off clay, she let it fall in the wilderness, and noble Enkidu was created. There was virtue in him of the god of war, of Ninurta himself. His body was rough, he had long hair like a woman's; it waved like the hair of Nisaba, the goddess of corn. His body was covered with matted hair like Samuqan's, the god of cattle. He was innocent of mankind; he knew nothing of cultivated land.

"Enkidu ate grass in the hills with the gazelle and lurked with wild beasts at the water-holes; he had joy of the water with the herds of wild game. But there was a trapper who met him one day face to face at the drinking-hole, for the wild game had entered his territory. On three days he met him face to face, and the trapper was frozen with fear. He went back to the house with the game that he had caught, and he was dumb, benumbed with terror.

"So the trapper set out on his journey to Uruk and addressed himself to Gilgamesh saying, 'A man unlike any other is roaming now in the pastures; he is as strong as a star from heaven and I am afraid to approach him. He helps the wild game to escape; he fills my pits and pulls up my traps.' Gilgamesh said, 'Trapper, go back, take with you a harlot, a child of pleasure.'

"Now the trapper returned, taking the harlot with him. After a three days' journey they came to the drinking-hole, and there they sat down; the harlot and the trapper sat facing one another and waited for the game to come. For the first day and the second day the two sat waiting, but on the third day the herds came; they came down to drink and Enkidu was with them.

"The trapper spoke to her: 'There he is . . . when he comes near uncover yourself and lie with him; teach him, the savage man, your woman's art, for when he murmurs love to you the wild beasts that shared his life in the hills will reject him.'

"She was not ashamed to take him, she made herself naked and welcomed his eagerness; as he lay on her murmuring love she taught him the woman's art. For six days and seven nights they lay together, for Enkidu had forgotten his home in the hills.

"'You are wise, Enkidu, and now you have become like a god. Why do you want to run wild with the beasts in the hills? Come with me. I will take you to strong-walled Uruk, to the blessed temple of Ishtar and of Anu, of love and of heaven: there Gilgamesh lives, who is very strong, and like a wild bull he lords it over men.'

"When she had spoken Enkidu was pleased; he longed for a comrade, for one who would understand his heart.

"'I will challenge him boldly, I will cry out aloud in Uruk, "I am the strongest here, I have come to challenge the old order, I am he who was born in the hills, I am he who is strongest of all."'

"She said, 'Let us go, and let him see your face . . . O Enkidu, you who love life, I will show you Gilgamesh, a man of many moods; you shall look at him well in his radiant manhood. His body is perfect in strength and maturity; he never rests by night or day. He is stronger than you, so leave your boasting. Shamash the glorious sun has given favors to Gilgamesh, and Anu of the heavens, and Enil, and Ea the wise has given him deep understanding. I tell you, even before you have left the wilderness, Gilgamesh will know in his dreams that you are coming.'

"Now Gilgamesh got up to tell his dream to his mother, Ninsun one of the wise gods. 'Mother, last night I had a dream. I was full of joy, the young heroes were round me and I walked through the night under the stars of the firmament, and one, a meteor of the stuff of Anu, fell down from heaven. I tried to lift it but it proved too heavy. All the people of Uruk came round to see it, the common people jostled and the nobles thronged to kiss its feet; and to me its attraction was like the love of a woman.'

"Then Ninsun, who is well-beloved and wise, said to Gilgamesh, 'This star of heaven which descended like a meteor from the sky; which you tried to lift, but found too heavy, when you tried to move it it would not budge, and so you brought it to my feet; I made it for you, a goad and spur, and you were drawn as though to a woman. This is the strong comrade, the one who brings help to his friend in his need.'

"So Gilamesh told his dreams; and the harlot retold them to Enkidu.

"And now she said to Enkidu, 'When I look at you you have become like a god.'

"She divided her clothing in two and with the one half she clothed him and with the other half herself; and holding his hand she led him like a child to the sheepfolds, into the shepherds' tents.

"Then the woman said, 'Enkidu, eat bread, it is the staff of life; drink the wine, it is the custom of the land.' So he ate till he was full and drank strong wine, seven goblets. He became merry, his heart exulted and his face shone. He rubbed down the matted hair of his body and anointed himself with oil. Enkidu had become a man; but when he had put on man's clothing he appeared like a bridegroom.

"He took arms to hunt the lion so that the shepherds could rest at night.

"He was merry living with the shepherds, till one day lifting his eyes he saw a man approaching. He said to the harlot, 'Woman, fetch that man here. Why has he come? I wish to know his name.' She went and called the man saying, 'Sir, where are you going on this weary journey?'

"The man answered, saying to Enkidu, 'Gilgamesh has gone into the marriage-house and shut out the people. He does strange things to Uruk, the city of great streets. At the roll of the drum work begins for the men, and work for the women. Gilgamesh the king is about to celebrate marriage with the Queen of Love, and he still demands to be first with the bride.'

"At these words Enkidu turned white in the face, [pale]. 'I will go to the place where Gilgamesh lords it over the people.'

"Now Enkidu strode in front and the woman followed behind.

"The people jostled; speaking of him they said, 'He is the spit of Gilgamesh,' 'He is shorter.' 'He is bigger of bone.' 'This is the one who was reared on the milk of wild beasts. He is the greatest of strength.' The men rejoiced: 'Now Gilgamesh has met his match.'

"Mighty Gilgamesh came on and Enkidu met him at the gate. He put out his foot and prevented Gilgamesh from entering the house, so they grappled, holding each other like bulls. They broke the doorposts and the walls shook. They snorted like bulls locked together. They shattered the doorposts and the walls shook. Gilgamesh bent his knee with his foot planted on the ground and with a turn Enkidu was thrown. Then immediately his fury died. When Enkidu was thrown he said to Gilgamesh, 'There is not another like you in the world.'

"So Enkidu and Gilgamesh embraced and their friendship was sealed."*

★ ☆ ★

And so Enkidu came to Gilgamesh for what was the beginning of many adventures for the two companions. The story is as fresh today as when it was first written almost 5,000 years ago. But it is more than a simple story. It contains within it clues to the ancient cosmos. I believe that in the comings and goings of Gilgamesh and Enkidu we are looking at the earliest written account of the sun and moon and the circle of the zodiac.

The moon takes 27 days to go once around the earth, starting and finishing at a given sign. The month of the phases is a little longer, 29½ days, because the sun also moves along the zodiac, though more slowly, and the moon requires the extra days in the moon-month to catch up. The orbit is tilted, and the moon appears to swing above and below the sun's path in the zodiac. The moon is a globe of dusty rock 2,160 miles in diameter at a distance of 238,860 miles. The only light the moon has is the sunlight in space and the much fainter light from the stars and planets. Its shape in the sky depends on the way the sunlight strikes it on any particular night. The first stargazers explained the changing shape by the story of Enkidu.

Enkidu is the moon and Gilgamesh is the sun. For 3 nights

* From *The Epic of Gilgamesh,* trans. N. K. Sandars. Chapter 1, "The coming of Enkidu." Copyright © N. K. Sandars 1960, 1964. Reprinted by permission of Penguin Books Ltd.

(Left) Last crescent. (Center) Last quarter. (Right) Full.

The old moon in the new moon's arms.

The woman in the moon.

(Left) First quarter. (Right) First crescent.

the hunter waits with the woman until Enkidu appears. These are the 3 nights of invisibility before the thin crescent of the new moon shows in the west. (We call the first crescent the "new" moon, but mathematical astronomers pin the new moon to the precise second when it is in line with the sun.) After he appears, Enkidu and the woman lie together for six days and seven nights and he becomes "like a god." These are the nights when the crescent grows in size and reaches the half-moon shape.

During these 7 nights there is an interesting visual phenomenon known as "the old moon in the new moon's arms." Sunlight bounces off the white clouds of the earth and illuminates the dark side of the moon. We see the globe glowing faintly and held in the cusps of the crescent. If my interpretation is correct, the Sumerians would say, "The woman is in the arms of Enkidu"—a 6-day, 7-night embrace.

You can see the profile of the woman with the naked eye at first quarter. European cultures speak of a "woman in the moon" at this time. Her dark hair falls in tresses to the right, formed by the lunar seas of tranquility, serenity, nectar and fertility. Of

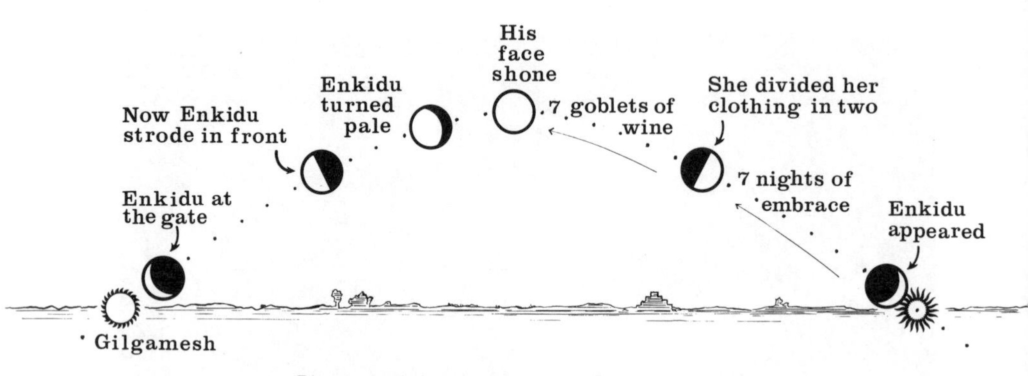

Phases of the moon in the Gilgamesh epic.

course, we know from the astronauts' visits that the so-called seas are flat, dry, dusty plains. Thus goes romance. What we now have named the Apennine mountain range makes the line of her forehead and nose, and two more seas make the dark eye and mouth. Seen from earth in the northern hemisphere the woman's profile glows faintly to the left of the crescent, and, since the moon moves to the left in the zodiac, the woman is leading Enkidu—"holding his hand, she led him like a child to the sheepfolds, into the shepherd's tents."

"She divided her clothing in two and with one half she clothed him and with the other half herself." That is a picturesque way of describing the sunlight creeping over the dark half of the moon. At first quarter the disc is divided in half. Seven nights later the disc is uniformly illuminated, or full. The 7 goblets of wine drunk by Enkidu in the shepherd's tent symbolize those 7 nights. At the end of the period he was "full," "his face shone," he rubbed down his "matted hair" and Enkidu had "become a man."

Visually, the change is equally dramatic. The full moon is nearly 10 times brighter than the half-moon phase, not just twice as bright. This is because the surface of the moon is rough and the sunlight is falling on it sideways at the half-moon phase. Good eating and 7 goblets of wine apparently accounted for this anomaly for the Sumerians.

When the moon shines as a bright round disc, the sea of showers appears on the left-hand side. The profile of the woman disappears and we see the familiar face of "the man in the moon." His two large eyes are dark seas and so is his small round nose and open mouth. The man in the moon even has a line of bushy eyebrows! The Sumerians called him Enkidu.

Why does the story talk about the rubbing down of matted hair? On the inside edge of the crescent the sunlight catches the jagged surface of mountains and craters, and to a person with average-to-good eyesight the line is rough. At full moon the sunlight falls vertically on the surface. The moon is uniformly bright. The edge of the disc is clean-cut. The hair has gone.

The moon stays this way for 3 nights, a bright orb challenging the sun and exposing a hunter as if it were day. After this the moon goes through all the phases again, but in reverse order. The illuminated side is now on the left and the faintly glowing side is on the right. If Enkidu is the old crescent he strides in front, as the poem says, and the woman follows behind.

All the while the moon has been moving step by step during the course of the month toward the sun, Gilgamesh. The glare of the sun blots out a whole sign of the zodiac. This is the sun-sign, or the horoscope sign of a person's birth. This is what the poem calls the marriage-house, and the sunlight "shuts out the people." The so-called doorposts are the morning and evening twilight zones. When Enkidu and Gilgamesh meet, the encounter will be entirely invisible, shut out from view.

The sun-sign was a mysterious part of the zodiac for the ancients. They could not see what took place in this sign; they could only speculate. Astrologers call it conjunction with the sun. The Sumerians knew all the planets had to pass through this zone of conjunction. There were no exceptions. Poetically it seems the sun had "first rights" in the marriage-house. One by one the planets disappeared in the light of the morning dawn and reappeared a few weeks later on the other side of the sun-sign in the evening twilight. Only on the rare occasion of a total eclipse of the sun would the ancients be privy to a view of the stars and planets located in that part of the sky.

As the moon is about to enter the sun-sign it shows the thin crescent of the old moon and is difficult to see in the dawn light: Enkidu turned white in the face saying, "I will go to the place where Gilgamesh lords it over the people." Enkidu is about to challenge Gilgamesh by entering the sun-sign.

The Queen of Love is mentioned at this point in the story,

and the king is about to "celebrate marriage" with her. The love goddess was the chief goddess of Babylonia and Assyria and represented fertility, pleasure and life-giving power. Like all the gods and spirits she lived in the heavens, and to the originators of the story, the Sumerians, she was Inanna the Lady of the Heavens. She was Ishtar to the Babylonians, Athtar to the Arabians and Astarte to the Phoenicians. The greatest female in the sky added power to Gilgamesh at this point by her presence in the bridal chamber.

Ishtar, Astarte, Athtar, whatever name a culture used, the love-fertility-mother-goddess always personified the planet Venus. This is the second planet out from the sun, the brightest planet in the sky, the one that outshines all the stars. Venus is called the morning star when it is on the right-hand side of the sun. From there it moves into the dawn light as it travels around the zodiac. From morning to morning Venus becomes more difficult to see and would appear to "pale" as Enkidu does as he approaches Gilgamesh.

There are some strange coincidences in English literature. For example, Jonathan Swift described the two small and unusual moons of Mars quite accurately in *Gulliver's Travels.* It is always a puzzle how he did so because these two moons were not discovered until a full 157 years later! Was he clairvoyant? Then again, Queen Victoria's poet laureate, Alfred Lord Tennyson, described Venus in his long poem "Maud":

> For a breeze of morning moves,
> And the planet of Love is on high,
> Beginning to faint in the light that she loves
> On a bed of daffodil sky,
> To faint in the light of the sun she loves,
> To faint in his light, and to die.

Tennyson is describing the fading of Venus in the dawn twilight as "she" approaches the male sun. "To die" meant to have intercourse in the vocabulary of medieval chivalry, and judging from the tone of the rest of the poem, Tennyson meant it that way. "Maud" scandalized the critics and they rated it with a flat Victorian "X," almost ending Tennyson's career. From the astronomical standpoint he described nothing more than the fading of the planet at dawn as it moves into conjunction with the sun. The similarity of his verse with the story on the Gilgamesh tablets is interesting because he wrote the poem in 1854, 35 years before Peters dug out the fragments. The Sumerian epic was translated in stages, with the story of the Flood coming first and the remainder being added in this century. Perhaps Tennyson was born with the idea in his subconscious, or perhaps the same visual phenomenon generated the same literary response in him as in the preliterate Sumerians.

One of the coincidences in astronomy is the way the size and distance of the sun and moon work out. The sun is much larger than the moon, but it is also much farther away. It just so happens that the discs of the sun and moon as seen from planet earth have the same size. This means that the moon can exactly blot out the light of the sun in a total eclipse. The outline of the moon slowly covers the sun like a blackened silver dollar sliding over a bright uncirculated one. This is a coincidence of space. It is also a coincidence of time, because it neatly fits into the time span of Homo sapiens. The moon used to be closer. The orbit is gradually enlarging, and 100,000 years from now the moon will have too small a disc to eclipse the sun. Earthlings will no longer be able to see the stars come out in the daytime. Eclipses which take place will be annular, with the sun showing a bright ring. In our time frame the distances are correct for perfect eclipses, and the sun gets eaten away as the moon moves in front of it. Just before totality the edge of the sun shines as a thin crescent, then the blackened dollar exactly fits over the bright one.

I believe that ancient peoples knew about this coincidence in size. The white moon was so different from the heat-giving sun, yet it was an exact counterpart to the sun. That is why the people in the streets of Uruk wonder and argue about the relative size of Gilgamesh and Enkidu. Enkidu is shorter, he is bigger of bone, they say. Finally the men pronounce the sun and moon to be equal—Gilgamesh has met his match, they say.

What happens in the Gilgamesh epic when the moon actually enters the sun-sign on the zodiac? There is a wrestling match and they grapple, holding each other like bulls. The metaphor refers to the horns of the old moon as it enters the sign and the horns of the sun during an eclipse. The wrestling match is the conjunction of the sun and moon when an eclipse might take place. Early cultures were apprehensive of this time of the month. If the moon passed above or below the sun, then no eclipse took place. If it passed exactly in front, at the node, there was a total eclipse of the sun. If it was slightly higher or lower the eclipse was partial and for an hour or so the sun would shine with a piece eaten out, looking like a crescent moon. The metaphor about the bulls refers to the cusps of the partially eclipsed sun and the crescent moon.

The Sumerians were probably taken by surprise whenever the sun was eclipsed. Events in the house of the sun were a mystery. But the Babylonians who came after them were able, by about the year 500 B.C., to calculate the position of the moon hour by hour even though it was hidden in the house of the sun. They could predict whether or not an eclipse would take place. They knew from long experience that the moon could be relied upon to reappear. Each month the moon passed by the sun during the 3 days of invisibility and safely left the marriage-house. A day or two after passing the sun it could be seen once again shining as a thin crescent in the evening sky. But in this earliest of stories we are reassured that the moon is a permanent part of the cosmos and the sun-moon conjunction is harmless: "with a turn Enkidu was thrown . . . and their friendship was sealed."

The modern Space-Age civilization knows the moon can be walked on as a barren, rocky globe. The ancients did not. Yet they came very close to modern knowledge. They had a goddess conceive an image, which was of the stuff of the firmament, and she made Enkidu by pinching off a ball of wet clay and letting it fall in the wilderness. It would certainly dry out if left in the desert. And it was so enormous that strong Gilgamesh could not budge it. When shaped by the goddess "the body was rough." Their picture of the moon was close to that of the modern astronauts.

The poem is a cycle like the phases of the moon and it can be told and retold. Judging by the widely scattered copies it was the favorite story of Middle Eastern civilizations for hundreds of years. It was at once an explanation, a description and an entertainment. It was a cohesive pattern for the transmission of information. It was reassurance that the moon and the sun would stay in their courses.

From the astronomical point of view the section where the

hunter becomes alarmed and goes to speak with Gilgamesh is a prologue to the cycle. From the time that Enkidu meets the woman by the drinking-hole the events flow in sequence, but before the meeting the events are somewhat isolated. This prologue as in all good stories is used to set up the characters of Enkidu and Gilgamesh. We hear how the trapper and Enkidu meet face to face for 3 days in a row and how Enkidu disturbs the wild animals and they escape. In another section the trapper says the wild game slips through his fingers. This could be the 3 nights of the dark of the moon when it was difficult to do night hunting. During those nights the moon is in the sun-sign and is face to face with the sun. Actually, although it cannot be seen, it is face to face with the sun at noon in the daytime sky. Alternatively it could be the hunter himself who is face to face with Enkidu, and this would be the 3 nights of full moon when the bright moonlight helps the wild game to see the pits and traps. The trapper takes 3 days for the journey, but we do not learn how long he takes to return. Since there is a lot of coming and going, one can take these incidents as setting the scene and providing an excuse for repeating the number "3," which is important for both the full and dark phases of the moon. On the other hand, if one insists on a literal fitting in of these numbers with the flow of the narrative, the extra days would have Enkidu leaving the shepherd's tent at last quarter, but even so, the allegories are the same.

I first learned about the Gilgamesh epic from the late Giorgio de Santillana. He had come over from M.I.T. to be a guest lecturer in my astronomy class at Boston University. He mentioned how the whole poem was "astronomical from A to Z." Afterward when I asked him for a clue he said, "You're the astronomer, not me." It was some time before I obtained a copy and read it through. Indeed, as he had said, there were the obvious clues. Over the next few years I went back to the story to dig deeper, and slowly this astronomical picture emerged. I telephoned his office to check on his ideas and was saddened to learn instead of his untimely death. But I think I might have found the interpretation he was hinting at.

Of course there is more to the poem than the chapter on the coming of Enkidu. It goes on for 10 more tablets with several more stories, and the cosmic thread runs through them all. Their first adventure as companions takes them to a dark forest which is the lair of the monster Humbaba, who is a dread enemy of both Gilgamesh and Enkidu. I take Humbaba to be the cause of eclipses, making the sun go dark at noon and the moon go a deep coppery red at night. Together they slay Humbaba, and forever after the two companions have "fire" on their faces. There is an intriguing phrase repeated several times in this adventure which is an astronomical clue. The line says how in 3 days the companions had walked as much as a journey of a month and 2 weeks. Now the moon at perigee, when it is closest to the earth and moving at its fastest, travels 44 degrees along the zodiac in a period of 3 days. The sun moves more slowly and takes a month and 2 weeks to cover the same distance (30 plus 14 days). Gilgamesh in his haste to kill the monster speeds up so as to travel in company with the moon.

On another adventure Gilgamesh and Enkidu kill the bull of heaven and Enkidu tears off the bull's right thigh—an explanation for why the star pattern of Taurus shows only the horns, head and shoulders of the bull.

Gilgamesh wants to live forever like the gods, and in another of the tablets he makes a long journey in search of the secret of everlasting life. This journey of his seems to be a complete circuit around the zodiac like the sun's, starting at midsummer in Leo, passing through Scorpio and returning to Leo. There are 3 seasons, each lasting 4 months of 120 days' duration. The seal of Gilgamesh unearthed at Uruk shows him with flying lions, a symbol of the sun. It is as though there were doubt in the Sumerian mind that the sun itself could live forever. They had many fears and insecurities. In this story sun-god Gilgamesh apparently does not live forever, but he is "renewed" when he gets back to the starting point of his journey.

Halfway along the journey Gilgamesh must follow the downward path of the sun, fighting dragons and beasts and going through a long, dark tunnel. He finds that the tunnel part of the zodiac is protected by a man-scorpion with no claws, only a tail. The star pattern of Scorpio in the sky has a well-marked tail glowing with the giant red star, Antares. But there are no claws. So the poem agrees with and explains the zodiac pattern around 3000–4000 B.C.

The epic of Gilgamesh is not the work of one person. It was added to and modified by the greatest brains of that story-telling culture as new insights and awareness of the cosmos came along. It is a collective work and its authors are anonymous. I believe the epic was used partly as a memory device to transmit astronomical knowledge and to explain the universe as it was known. It also has other qualities. It is legend and allegory, philosophy and literature, inspirational and religious. These are humbling qualities to be found on the first tablets ever to be inscribed. We learn about the concerns and fears of those cultures, their way of life and their pleasures. Yet we earthlings of the 20th century are not so far removed. Gilgamesh in the tunnel episode craves to be permitted to live forever and to have undimmed glory and renown. One of the gods of the cosmos speaks to him from on high, saying:

"'Gilgamesh, where are you hurrying to? You will never find that life for which you are looking. When the gods created man they allotted to him death, but life they retained in their own keeping. As for you, Gilgamesh, fill your belly with good things; day and night, night and day, dance and be merry, feast and rejoice. Let your clothes be fresh, bathe yourself in water, cherish the little child that holds your hand, and make your wife happy in your embrace; for this too is the lot of man.'"

FOUR
The Morning Star

In mindstep 1 the cosmos was viewed like a planetarium show without the machine humming and controlling things from the center. The sun, moon and planets had an occult power of their own. They moved with their own force, willfully and independently. The slightest changes in the movement of the gods could be the signal or cause for changes on earth below. The panorama of the machineless planetarium was watched by the earthling audience carefully, apprehensively and meticulously.

The ancient stargazers saw two inner planets, Mercury and Venus, which traveled first to the left and then to the right of the sun. When they were on the left-hand side of the sun they shone as bright evening stars, and on the right they showed as morning stars. As a morning star, Venus was the brightest star-like object in the sky, outshining Jupiter and Sirius, and second only to the moon. Mercury was a threatening, mysterious interloper, lurking in the glare of the sun. To the ancient Greeks, Mercury was a fleet-winged messenger because of its rapid movements in the orbit nearest to the sun.

There were 3 other star-like gods—red Mars, white Jupiter and yellow Saturn. The ancients watched these planets go from sign to sign in the band of the zodiac, sometimes above the ecliptic, sometimes below. As outer planets, Mars, Jupiter and Saturn

NOTE: This chapter is based on work done in Mexico, 1973.

could take up a position opposite to the sun, when they would dominate the midnight sky. They each made a short reversal in their motion at opposition, falling back in the zodiacal sign—retrograde motion—now known to be an optical illusion caused by planet earth (on the inside track) overtaking the slower outer planet. Soothsayers made predictions as the planets moved from sign to sign: "Beware when Mars the god of war is retrograding, that blood-red midnight star is the color of battle, his retrograding shows an evil plan for mortals . . . "

Did the ancients know that Jupiter is the largest and most massive of the planets? There is no way to tell by eyesight alone. It is not as bright as Venus, nor does it have the longest period, yet those observers singled out Jupiter as the major god, the most beneficent, the most influential—the Jove of the Romans. Saturn on the other hand brought misfortune. As the slowest-moving planet it was associated with aging and death, at least in Indo-European cultures. God Saturn was said to be on the boundary of heaven; he was the keeper of the frontier. The ancients were right in this. Saturn is the farthest planet visible to ordinary unaided vision.

With sharp eyesight and under exceptional conditions it is just possible to see Uranus, the 7th planet out from the sun. We know from calculations that when Uranus is closest to the earth and set against the darkest parts of the zodiac on a clear moonless night, it shines feebly just above the visual threshold. But the ancients did not see it. Or if they did, they did not make a record. For them Uranus was beyond reach, unknown and unnamed. Slow Saturn marked the planetary abyss. Nothing was thought to exist beyond it except the stars.

There were two disc-gods, the sun and moon. One was hot and radiant, the giver of life. The other was pale white, cold and changeable in shape. But the moon was also connected with the life force—the phases kept in step with (controlled?) the human fertility cycle, and moonlight was supposed to germinate seeds and make things grow. In many ways the moon-god was more powerful than the sun. It could change its shape, and it could shine both day and night, while the sun-god was condemned to go down into the underworld each night at sunset.

Sun and moon were included with the planets, making a total of 7 astro-gods who played out events in the great circular arena of the zodiac. In the ancient world, 7 came to be a cabala number—magical, celestial, a number of the cosmos, a direct product of mindstep 1. We have it today in the number of days in the week, the seven deadly sins, and "lucky 7." Maybe even the sales of the soft-drink 7-Up have a lift from the cabala connection!

Babylonian clay tablets give us a clue as to how the Mesopotamians figured out the movement of a planet along the zodiac. Other cultures before them may have had similar knowledge, but without a written record we cannot tell—they didn't publish and they perished. The Babylonians tracked the planets by adding a new number to the old one, step by step. It was a laborious process, but it worked. They had a compulsion for knowing the positions of the sun, moon and planets.

Each planet moves according to a definite timetable called the synodic period. The planet acts out a shadow-play with the sun, and when the final move has been done, the story begins all over again. The synodic period is a sort of magic number for a planet. Once it is known, the movements, changes and aspects are predictable.

Venus was watched more carefully than any other planet. I do not support Immanuel Velikovsky's theory about Venus coming as a comet from out of Jupiter and toppling the earth's axis. My reasoning is based on celestial mechanics—the gravitational forces would have been insufficient to do this, and no other recognizable physical forces have been proposed—but Velikovsky's research into the myths, legends and ancient records certainly shows that planet Venus was regarded as a most important object in those times.

Astronomically, as an inner planet Venus circles the sun inside the earth's orbit. It passes between the earth and the sun, when it is lost in the glare for about 14 days. This is known as the short encounter (inferior conjunction), and at this time Venus is closest to the earth. Later on in the cycle the planet passes behind the sun on the far side of its orbit—the long encounter—and is lost to view for almost 3 months. At this time it is farthest from the earth (superior conjunction).

The ancients figured the Venus cycle, the synodic period, to be 584 days long—the time taken by the planet to move once around the sun as seen from the moving earth. They knew Venus showed as the bright star of the evening when it was to the left of

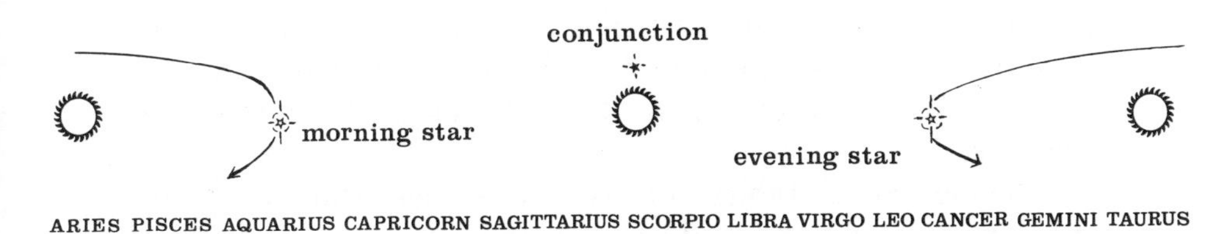

Manifestations of Venus.

the sun, and as Venus morning star when it was to the right. Ancient Greeks called Venus, Phosphoros, "the bringer of light," when they saw it in the morning sky, and Hesperos, "the western one," when it showed at sunset. But even with two names they knew it was one and the same planet. These apparitions were just different events in a long 584-day story. For mid-latitude countries it went like this:

Venus Manifestation	Length (days)	AGE (from Superior Conjunction)
evening star	246	39
short encounter	14	285
morning star	246	299
long encounter	78	545

The Babylonians (B.C.) and the Maya (A.D.) were both in mindstep 1, and they kept track of Venus in a very straightforward way. They took a starting date, added the length of the manifestation to it, and came up with the new date. For the Maya an extra glyph was painted in the pages of the codex. For the Babylonians, line by line, tablet by tablet, planet Venus was pressed into the clay. Only the experts did this. They were the calculators, the keepers of the record. On some of the tablets there is a classification line saying in effect: "Secret, not to be divulged."

For those today who are mathematically minded the same tracking of Venus and the planets can be done with a home computer or pocket calculator. I have worked out a few programs and "cookbook recipes" for getting to the right answer. They are in the Appendix, and these programs have been put together with the advantage of modern parameters and equations, but deep down inside the silicon chips the computer is carrying out operations very similar to those of the tablets or codex. Programs and computed quantities are written for convenience in capitals—AGE, ASTRODATE, LONGITUDE, etc. These terms are explained in the Appendix.

The electronic "tablet" has two advantages that the ancient scribes would have relished—speed and accuracy. And it goes without saying that the keyboard is far more convenient than a stylus and wet clay. The machine is guaranteed not to make errors in addition or subtraction as the Babylonian scribes sometimes did. Troubles come only from bugs (errors of logic) in the program, or the unfortunate pressing of the wrong button by the operator. There should be no bugs in the Appendix programs—these were developed carefully over a period of several years and were thoroughly tested out by a group at the Smithsonian Institution—and pressing the wrong key can be avoided by checking step by step, line by line, in the worked examples.

For those who are not mathematically minded the computer programs and anything in SMALL CAPITALS can safely be ignored, bearing in mind that planets do indeed move by numbers, and that silicon chips can now do all the mathematical work. But before relegating math to the Appendix there is one remarkable thing that the ancients noticed: planet Venus traces out a 5-pointed star as it travels in the zodiac.

Suppose Venus is an evening star in 1985. It will be an evening star again after 584 days, or 1.6 years later. Adding 1.6 to 1985 in the calculator gives 1986.6 (the decimal fraction 0.6 means that the date is the 220th day after January 1, or August 8). By adding 1.6 for each step, the dates, as shown by the decimal fraction, make a number pattern:

1985.0	1986.6	1988.2	1989.8	1991.4
1993.0	1994.6	1996.2	1997.8	1999.4
2001.0	2002.6	2004.2	2005.8	2007.4 . . . etc.

After 8 earth-years the fraction repeats; the year 2001 A.D. is 2 cycles beyond 1985. This means that Venus is an evening star again during the same season of the year (winter), in the same sign of the zodiac (longitude 327), at the same spot in the heavens. The planet takes up 5 equally spaced positions in the band of the zodiac between 1985 and 1993, and repeats them between

1993 and 2001. Every 8 years, exact to within a day or so, Venus traces out the tips of an enormous 5-pointed star with the earth-observer at the center. The pattern repeats itself again and again, and the tips of the Venus-star are fixed against the stars of the zodiac constellations.

This was nothing short of a miracle for the ancients. In all the confusion of planetary movements and the endless calculations, here was a simple, reliable pattern. The 5-pointed star became an important symbol. The Greeks wrote about it and philosophized, the Babylonians watched the planet turn in the zodiac in tune with their clay tablet numbers, and the Egyptians painted the 5-pointed star on the ceilings of their tombs.

Venus is tied to the sun by gravity, like a ball on the end of a string. It can never be more than two zodiac signs away. If the sun is in Aries, then Venus as an evening star is in Taurus or Gemini. If the sun sign is Gemini Venus can go no farther east than Leo, and so forth. The tips of the 5-pointed star are traced out as the sun carries the planet around the full circle of the zodiac.

Mercury is more constrained than Venus because its orbit is smaller. It is never more than one zodiac sign away from the sun. Mercury's manifestations repeat after a period of 116 days. It traces out a 3-pointed star on the zodiac, but it is not as accurate as Venus's and was never noted in antiquity.

Numbers and words are processed in the left hemisphere of the brain, art and visions in the right. This is the rule for modern humans, and the same probably held true for peoples of the past. A few were selected, or selected themselves, to do the numerical work. Most others preferred the enjoyments of the imaginative side of the brain, lyrical rather than logical, inspirational rather than pedantic. The epic of Gilgamesh is an example of a flight of fancy with astronomical numbers scattered in it. But sometimes the dominance of the right hemisphere made those people fearful rather than factual, and slaves of their own creations.

The pre-Columbian Indians of Central America were mathematically minded and imaginatively artistic. They watched and noted planet Venus for centuries. The Maya in Yucatan knew about the 4 manifestations of evening star, short encounter, morning star, long encounter, and they gave the lengths in days as 250–8–236–90. These figures are a little different from those at the beginning of this chapter, which apply to the higher latitudes. Maybe the Maya could see Venus for a longer time in the dawn glow. They were near the equator and dawn comes on suddenly in that latitude. This would account for their making the short encounter 8 days instead of 14. But there is no way of explaining why the evening and morning segments are of different lengths, 236 and 250. The astronomy actually dictates equal periods of 246. Maybe there was some nonastronomical reason which made them take out 14 days from the morning star segment and add it to the fourth segment of the "long encounter." Other civilizations of the area divided the segments a little differently from the Maya. The Borgia Codex gives the divisions as 252–12–243–77, which is closer to the northern-hemisphere values. But the total of the numbers always came out to be 584, the magic number for Venus.

It wasn't easy to come up with the correct value, because the synodic period of Venus is not really constant. It varies from about 579 to 589 days, because of gravitational disturbances caused by other planets and because of the elliptical shape of the orbit of the earth. The correct number could only be arrived at by careful observation over the generations, by counting off and remembering the days and the years.

Pre-Columbian cultures linked Venus to the god Quetzalcoatl and that god's legend. The Aztecs and the civilizations before them—Toltec, Maya, Totonac—had different versions of the Quetzalcoatl story, but each was a version of one main theme. The Aztecs clocked Venus, and they told their legend as the planet moved, and the legend ruled their lives.

Venus for the Aztecs and the cultures of the New World was not the goddess of love. Venus was a male. When "he" shone in the twilight, then Quetzalcoatl walked the earth as a human male. Quetzalcoatl was a wonderful, loving, intelligent leader, but only when Venus was an evening star. In the dawn twilight Venus was a dread sky object, Quetzalcoatl the plumed serpent god of vengeance—Venus morning star.

"Gilgamesh" is the oldest story of the Old World; "Quetzalcoatl" is thought to be the oldest story of the New. It shows up in all the cultures of Central America and in the Andean folklore of South America. Like the Gilgamesh epic it was probably told as a story before the invention of New World writing. By the 14th century A.D. it was a meme in the minds of the Aztecs, the overlords of the region. There is astronomy in the story, and the legend together with the astronomy was to be their downfall. Although anthropologists are still working out the details, "The Legend of Quetzalcoatl," as I paraphrase it, goes like this:

★ ☆ ★

"There were five goddesses of the moon, all sisters, and they went to the top of a hill one day to do penance for their sins. The first one was named Coatlicue, and she wore a petticoat fringed with serpents. The sun was her husband and he was made bright and powerful by her.

"Coatlicue stood on the hill and was made pregnant. Life stirred within her. Some say it was when a colored feather touched her bosom. Others say the sun put an emerald in her mouth and that was how it happened. In due course she swelled and gave birth to the great god who was called Quetzalcoatl. His name meant 'Serpent of Plumes and Feathers.' He had a twin who lived in the spirit world and who sometimes took the form of a dog.

"But Coatlicue already had a large family. Some say she had 400 sons, all warriors. Others say the number was 4,000. These sons shouted: 'Who is it who has made our mother pregnant!' When they were told it was the sun they were filled with hatred for him. They made war on the sun-god. The 400 brothers made war and killed the sun and buried his body in the sand.

"A vulture flew to Quetzalcoatl who was in the shape of a 9-year-old boy and told him of the murder. A coyote, an eagle and a wolf helped Quetzalcoatl find the body, and an army of moles took him down to the dark underworld to reach the bones. To avenge his father's death he battled and killed all of the 400 warriors.

"The gods transformed Quetzalcoatl into the shape of a handsome man, broad of brow, light-skinned with large eyes and a fair beard. His face was smeared with soot to show he belonged to the dark night sky. He wore a necklace of seashells and on his back hung a long-tailed quetzal bird, the beautiful green bird of the jungle treetops. As a man he was given strength and a pure heart. He foreswore lust and pleasures, and he renounced human sacrifice. He lived in a temple with 4 rooms facing east and west, south and north, lined with gold, emeralds, seashells and silver. From a hill in Tollan his voice could be heard for a distance of 10 leagues. He turned himself into a black ant and stole grain from the red ants. When he walked the earth as a man he taught the people how to grow corn and to make pottery. He was peaceful, he resisted all temptations and he was filled with a love for all things. At midnight he would go to the banks of a stream with his sister to bathe and to pray.

"But time changed him until his face was not human at all. The eyes were sunken, the skin pale and wrinkled, the eyelids inflamed. When he looked in a mirror he saw his weakness and he said: 'If the people see me they will shout, "He is weak, he is pale!" and they will kill me.' So Quetzalcoatl went into retreat.

"A deceitful enemy came upon him and spoke with him. The enemy dressed him in a garment of fine feathers. He painted his lips red like a woman's, he painted his forehead yellow like the sun and put on a beard made of beautiful bird plumes. When it was done the enemy showed him the mirror and Quetzalcoatl was pleased.

"Three enemies took him to their house. They brought him the intoxicating juice of the maguey plant and said, 'Drink!' but he refused, remembering his vows. Quetzalcoatl dipped his fingers in the goblet and it tasted good. So he drank 5 full goblets. With wine he grew merry, and his servants too. The company no longer led a pure life, and they became very cruel. Quetzalcoatl called for his sister and she too drank 5 goblets and became merry with wine. There was music and feasting in the house and all sensual pleasures. Brother and sister lay with each other and sinned.

"Time passed and Quetzalcoatl sank to total despair. He felt he was no longer a god, he was not the exalted son of the goddess with the serpent petticoat. He sought a judgment from the oracles. The signs were all against him. He decided to do penance for his sins. He ordered a stone casket to be made, and he lay in it as if dead, some say for 4 days, some say for 8. He buried his gold and all his treasures, and he commanded his people to march with him to the east, to the seashore, saying: 'The sun-god calls me to him, I must go.' With his followers he wandered to the land of the sun. 'Do not turn back,' he called, 'keep a firm step. Something you will achieve.' Along the way demons stripped him of his arts, of agriculture, of jewelry-making, of writing and poetry.

"When he arrived at the seashore Quetzalcoatl dressed himself in the royal feathered robe and covered his face with a turquoise mask. A yellow rectangle was painted on his forehead. Some say he then stepped onto a magic raft propelled by serpents and floated out to sea, telling the people he would one day return as a god and noble king.

"Others say he collected wood and built a large fire by the seashore. He threw himself upon it until he was consumed in the heat. The ashes of his body straightaway turned into birds of paradise. The heart rose up from the flames into the morning sky, and the blood became Venus-Quetzalcoatl, the bright morning star. From his throne in heaven Venus shoots down arrows at the world, at warriors and at priests and emperors and all people."

★ ☆ ★

An interesting story, full of numbers and journeys and time. Anthropological experts say the 400 (4,000?) warriors are the number of stars visible in the sky; they say the underworld is where the sun goes at night, and Quetzalcoatl is the planet Venus. Some say Mercury, Mars and the moon are there, too. So it might be useful to take an astronomical look . . .

Why are there 5 sisters? Probably these are the tips of the 5-pointed star, the places, the stations in the zodiac belt where Venus first appears as an evening star. The 5 sisters are placed in a ring around the earth to give birth to Quetzalcoatl as the sun mates with them. Perhaps the 5 goblets are another reminder of the Venus stations.

When the sun sinks in the west the stars begin to shine. The stars of the night are victors of the lord of the day. So at evening the "warriors" have killed the sun and buried his body below the western horizon.

It is just the reverse when Venus is a morning star. The stars fade at dawn and brilliant Venus is the last to disappear. At maximum brilliance, Venus can be seen in daylight, shining with the sun. Quetzalcoatl is now victor over the warriors. It is approximately 260 days from evening star to morning star, so Quetzalcoatl appears at dawn and kills the warriors in the 9th month after his birth. This could be the meaning of the 9-year-old boy. For the Venus-god one lunar month is one year of aging.

After Venus morning star has killed the warriors he digs up his father's bones in the underworld. Astronomically that must be the time of the long encounter when Venus is invisible, as it sets each evening with the sun in the burial ground of the western horizon. This is a long period of time, 78 days.

We may be dealing with overlapping versions of the legend, but if Quetzalcoatl takes his vows of purity during this segment he will begin his walk on earth as Venus emerges for a second time above the western horizon as the evening star. Venus has moved from one Coatlicue sister to the next, from tip 1 to tip 2 of the star-shape in the zodiac. When he walks the earth, Quetzalcoatl is depicted as truly a mortal, but anthropologists say the smearing of his face with soot is significant—it shows he came down from the black night sky. The golden reign has come. It will last so long as Venus is an evening star.

But there are forebodings. These show astronomically as the planet reaches maximum elongation and retraces the zodiac path to the sun. The planet is fading. It no longer is the most brilliant object in the heavens. Quetzalcoatl on the earth is becoming ugly and resorts to wearing a mask and painting his lips.

The temptation of Quetzalcoatl probably takes place during the 14 days of the short encounter. At this time the planet disappears in the west and reappears in the east. The sinning, once begun, probably continues through another segment of the morning star aspect because judgment and penance take place in the second visit to the underworld. Down there he is found guilty and he is condemned to wander as an evening star. This sad wandering ends at the third short encounter with the sun. This time Quatzalcoatl is consumed by the fire and his heart rises to be transformed into the morning star. This last aspect of Venus as the spirit of the morning star was the most terrible to the Aztecs. Quetzalcoatl took vengeance on all mortals below and could be appeased only by human sacrifice.

Some anthropologists describe the Aztecs as a thoroughly guilt-ridden society. There was no escape for them from the dread of the Venus cycle—bad must always overcome good, a perfect god must fall in sin. The Venus orbit in the zodiac required from earthlings below a regime of penance, of vengeance and appeasement. Beware when Quetzalcoatl takes on the shape of Venus morning star! Evil has come, sacrifice is needed! The Aztecs tore out thousands of hearts from slaves and captured warriors to feed the Quetzalcoatl serpent and his father, the sun.

Some groups of Aztecs were against sacrifice, but those in power were for it. Even Quetzalcoatl himself was supposed to fall to the temptation of killing by sacrifice at that phase of the cycle when he walked the earth as the good god. It is difficult today to comprehend. Perhaps we never shall. Some psychologists have gone so far as to blame the cruelty on poor diet. In colonial Massachusetts, the hallucinations that spurred the Salem witch hunt have been traced to a certain fungus on the grain. In Central America the staple food was corn. According to some food experts a monotonous corn diet is deficient in the natural tranquilizer tryptophan. Corn without meat or vegetable protein is a cause of hyperactivity and violence, they say. This may or may not be so, but it could explain the unending cycle of excesses followed by penance in those years before the Spaniards arrived.

The Totonac kingdom was one of the main cultures of pre-Columbian America before the Aztecs. The Totonacs thrived in the classical period around 700 A.D. and were related through language to the Maya, but by 1500 A.D. they were dominated by the Aztecs. Their capital was at Tajin, 25 miles from the east coast of Mexico. It was a large complex of temples and pyramids, and the ruins can still be seen today.

They had built a great ball court in one of the plazas around 1000 A.D. The ball game was important in all pre-Columbian cultures, not so much for pleasure but for religion. Two teams hit a hard rubber ball and scored when the ball went through a small ring carved into the stone. They could not use their hands. The ball was deflected from their bodies. The backward and forward movement of the ball was symbolic of the movement of the planets—a goal was conjunction (mating) with the sun. But the game was not a sport in the modern sense of the word. The ball court was between walls, a sort of underworld. The team players were being tested, judged by the gods by trial of strength. The players were warrior stock and fierce. The evidence seems to suggest that the losing team was sometimes condemned to death and sacrificed at the temple.

Some time ago, 4 stone panels were discovered at the ball court of Tajin. I was present at a conference in Mexico City in 1975 when anthropologists discussed the carved-stone art excitedly. They were sure it was the legend of Quetzalcoatl as known to the Totonacs before the time of the Aztecs, and what was more, there were numbers hidden in the glyphs. Astronomy! I listened as the experts read the meaning of the pictures . . .

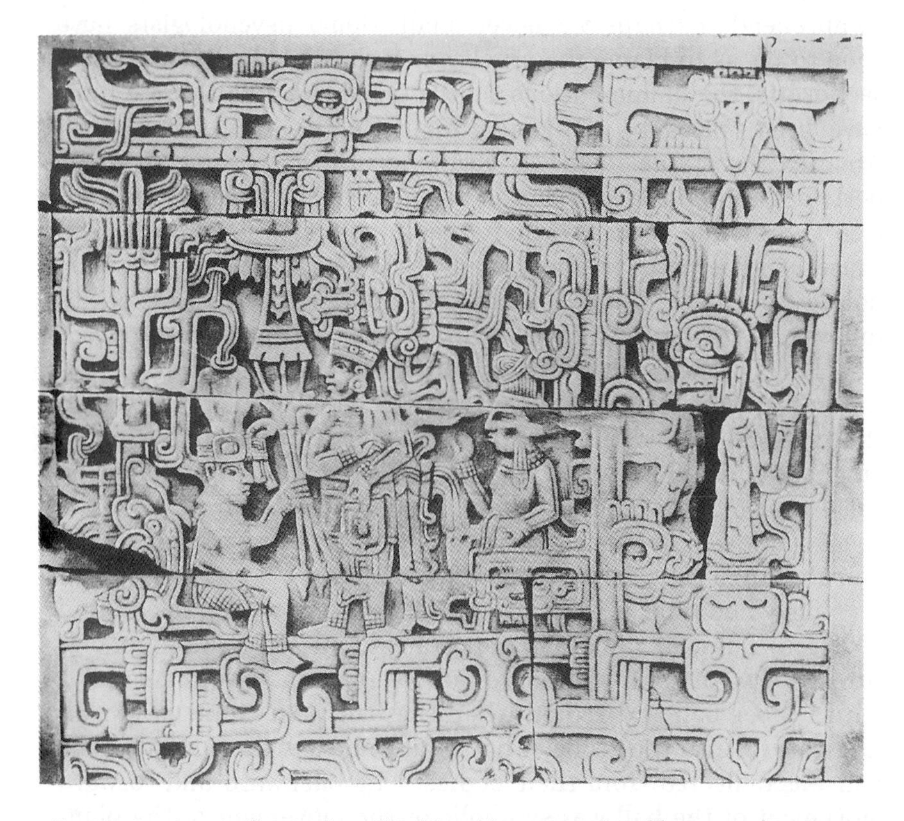

Panel 1 *shows Quetzalcoatl taking his vows of celibacy and peace as he starts his walk on earth. He sits on the left and the god standing in front of him and accepting the vows is his father, the sun, who has probably been brought to life after the digging up of the bones. Behind the sun is a planet-god. It may be Mercury, or Venus's supposed twin, in the underworld. On the far right there is a skeleton. The arrows in Quetzalcoatl's hand represent the male organ because they are connected to the loincloth above. The arrows are turned down to symbolize celibacy.*

Panel 2 *shows Quetzalcoatl as he starts his sinning. The hero-god is falling to earthly temptation, breaking his vow of celibacy and sealing his fate as the legend foretells. He is lying on a couch with his sexual partner on top dressed in the form of a bird. Looked at through Totonac eyes the scene was gravely lascivious because musicians played at Quetzalcoatl's head and feet. Enjoyment of music, like intercourse and drinking alcohol, was bad.*

The Totonac story was carved in stone long before the rise to power of the Aztecs. It is not the full legend, lacking the episode of the search for the sun-god's bones and the battle with the warriors—the 400 bright stars of the constellations or the 4,000 stars visible in the whole sky. Anthropologically the story is less cruel than the Aztec version because the image of Venus morning star as a killer star is not there. Instead, the good god himself goes to be sacrificed for sins on earth, and we can picture the

Panel 3 *shows Quetzalcoatl being sent off on his journey of penance to the sea. He has been judged and found guilty for his lapse of behavior down on earth. The judge is the sun, and a scribe sits on the left with a codex to record the verdict. The scene is in a ball court. The scribe sits on one wall, a dog-headed figure on the other. Mercury, or Venus's twin? Note the twisted strands above the round goal. In the ball game the players had to hit a rubber ball into a small hole. The strands show the movement of the ball and also symbolize the swinging orbit of Venus east and west of the sun. The scene is taking place in the underworld, so the anthropologists said, because the frieze at the base has no feathers. According to the words of the legend, Quetzalcoatl is condemned and starts his wandering as an evening star, benign and weakened, on his journey to the sea.*

Panel 4 *shows his death by sacrifice. In the Totonac picture version of the legend he does not burn to ashes, nor does he sail out to sea on a raft propelled by wriggling serpents. Here we see his heart being cut out by the sun-god, or a priest of the sun-god. There is a Totonac "diving god" with legs and arms hanging down from the zodiac map connected to the heart by a flame. The noble heart will rise into the cool dawn light to appear as Venus morning star . . .*

Totonacs at that time gazing up at the brilliant morning star as if it were a beautiful, green-plumed quetzal bird of paradise, a god who would return again according to the Venus cycle to walk the earth as a man.

Now for the astronomy. Today we live in a much later mindstep and the Totonac astronomy is quite different from modern ideas and theories. It is necessary to try to see things through the eyes of mindstep 1. The process is like future shock, but in reverse.

The first shock to the modern eye is the way the Totonacs depicted the sun, moon and Venus. Glyph experts say Venus is that indescribable shape in the maps with trailing flames or quetzal-bird plumes; the sun is the square buckle (?), as opposed to our notion of a disk; and the moon has no fewer than 4 glyphs, 2 long faces or vulvas, one marked with a dot, and 2 sectored designs, one left-handed, the other right-handed. Did the Totonacs see the sun, moon and Venus this way when they looked up, or was their art abstract symbolism? The second shock is the rectangle above each panel. Those complicated glyphs are a map of the zodiac at the instant of the picture in the panel below. Third shock: the narrow line separating map and panel is a moon clock. Each dot counts off the lunation (29.53 days), the tick of the cosmic clock. Fourth shock: panel and sky are connected on a one-to-one basis. Quetzalcoatl and the sun-god play out the scene directly below their personal astro-glyph. (There is a skeleton-like character who appears in each panel under the moon glyphs, but at the time of this writing anthropologists have not been able to determine the role of this "moon" character in the legend.) Fifth shock: it is not art for art's sake. The dots, feathers and decorative markings have numerical meaning; art and science are intertwined; the workings of the left and right hemispheres of the brain are combined!

I worked with the anthropologists in deciphering the carving and this was the resulting astronomical interpretation:

MAP 1

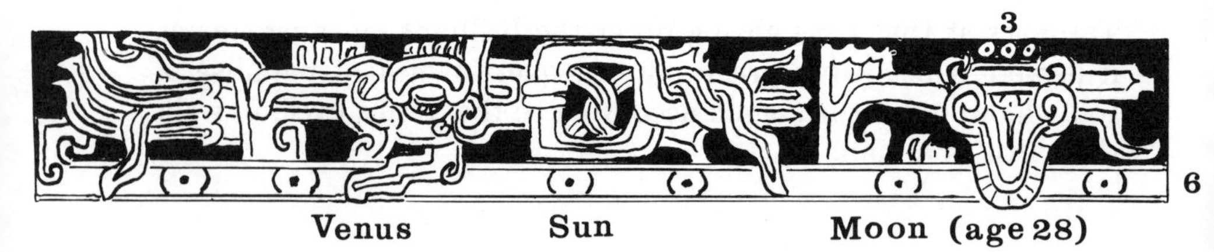

Here Venus is an evening star emerging from conjunction on the left side of the sun. This is the beginning of the manifestation of evening star which lasts for about 246 days. The moon is to the right of the sun, probably last crescent. The glyph is a face-on picture with a small spot in the center, which could be the faint shape of the old moon in the crescent illuminated by earth-light. One anthropologist speculated that the spot could be the emerald in the mouth of Coatlicue. A fertility sign would certainly be appropriate here because it is the time of the birth of Venus-Quetzalcoatl, immediately following conjunction with the husband-sun. Notice how in the panel below, Quetzalcoatl stands to the left of the sun-god, reflecting the configuration in the zodiac.

MAP 2

Now the evening star manifestation is over, because the Venus glyph and the sun's buckle are intertwined. Venus is in conjunction with the sun; it is the short encounter of about 14 days. There is an encounter of the flesh in the panel below—Quetzalcoatl's sexual partner is on top.

The moon has moved to the left of the sun and is shown as the right-facing sectored glyph. Probably this is the half-moon condition at the first quarter, the 9th day of the lunation counting from the last crescent, map 1.

The moon dots below map 1 show how many months have gone by between birth and sinning. There are 9 dots—6 in the narrow strip and 3 extras carried on the top of the moon glyph in the zodiac map. These carvings were made before the New World invention of zero, so the dots have to be counted "1st moon, 2nd moon, 3rd moon," etc. Map 2 is therefore 8 lunations and 8 days after the start of the story. It is day 244 in the Totonac version of the Venus cycle ($8 \times 29.53 + 8 = 244.24$), a good agreement with the astronomical value of 246.

MAP 3

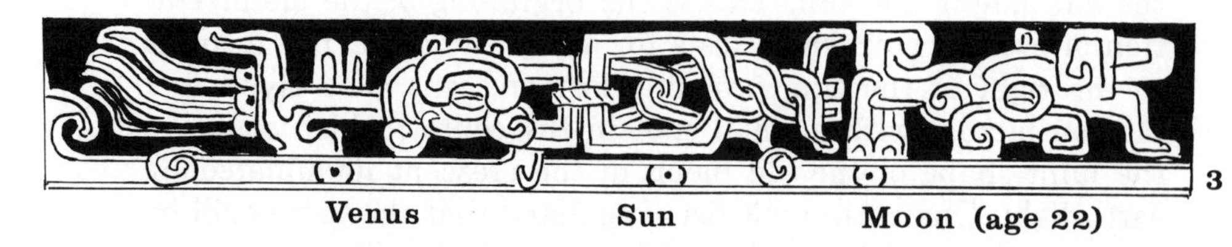

Venus and the sun repeat the alignment of map 1 with Venus showing on the left at the beginning of the evening star manifestation. Because of this, 584 days, more or less, must have passed. Is this number in the glyphs? The moon is shown as a sectored glyph facing left and is placed on the right-hand side of the sun glyph. The phase is probably half moon, last quarter, the 24th day of the month reckoning from last crescent. There are 6 moon dots on panel 1, 8 on panel 2, and 3 on panel 3—a total of 20. So we are in the 20th moon on the 24th day. The number of elapsed days is $19 \times 29.53 + 23 = 584.07$, the Venus period as expected from the astronomy.

Before commencing his long journey of penance, Quetzalcoatl has stayed in the underworld (superior conjunction with the sun) for about 3 months. This is clocked off by the 3 dots on the line under map 3.

MAP 4

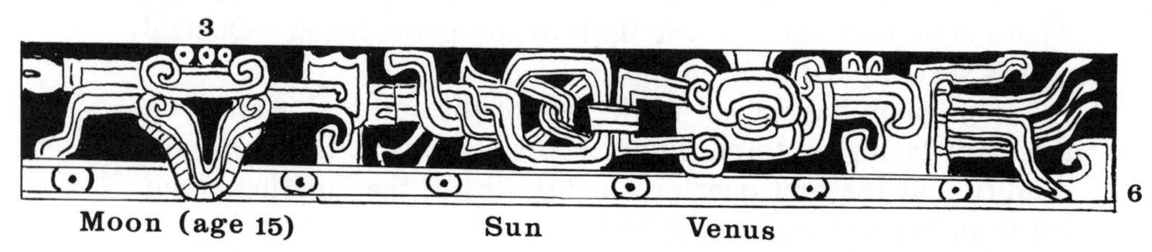

This is the only map that shows the Venus glyph to the right of the sun, the fatal morning star. According to the moon clock it takes place in the 29th lunation. The day of the month is shown by the phase of the moon glyph. It is a full-face presentation with the moon to the left of the sun. At first glance it is the same as the glyph in map 1, but there is a difference—it does not have the dot (emerald seed?) in the center. The astronomy suggests that map 4 shows the full moon, the 17th day from last crescent. If the emerald seed is an indicator of pregnancy and growth, lack of it could indicate barrenness—the moon has reached its waning state, it will not grow any bigger. If so the Totonac timing of the appearance of Venus morning star is the 17th day of the 29th moon, or 843 days from the start at panel 1.

For pre-Columbians, of course, everything was a cycle, the end was the beginning. After emerging in the dawn twilight, Venus would grow brighter, climbing day by day in the heavens until it reached its maximum brilliance, shining in broad daylight, unique among the stars. Then the planet would retrace its path to the sun, and 324 days later the viewer in the ball court panorama would be back again at the starting point, panel 1.

The Totonac Venus divisions can be figured out with a little algebra. The morning star appearance next after panel 2 is not marked in the story, but it would be at day 259 (843 − 584). The short encounter is 15 days in length (259 − 244). If the morning star manifestation lasted 244 days, equal to the evening star segment, then the long encounter is 81 days (584 − 244 − 15 − 244). The Totonac divisions of the Venus cycle were therefore 244-15-244-81, closer to the correct astronomy than the Aztecs' or the Mayans' and more knowledgeable because the moon's phases were woven into the sequence.

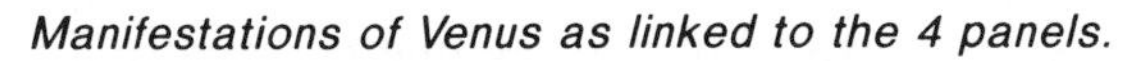

Manifestations of Venus as linked to the 4 panels.

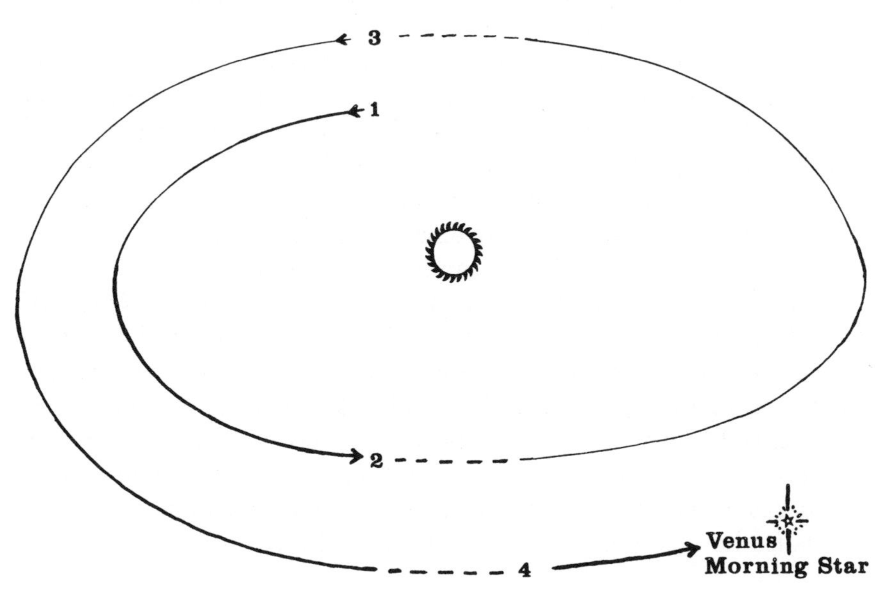

Not all of the features in the carving at Tajin have been decoded, and more work must be done. For example, what is the figure sitting under the moon glyph in all four panels? Larger than life, this mysterious skeleton is so far unidentified. And then what of all the other dots, feathers and markings in the pictures? Could these also carry astronomical information such as the periods of Mercury, Mars or Jupiter? At the moment we can only speculate. One suggestion concerns the upside-down god diving to earth from the Venus glyph in panel 4. This might be the soul of Quetzalcoatl, establishing a zodiac-earth connection. And why is the Venus glyph sectored? Could they have known it showed phases like the moon??

What would happen if we could go back to mindstep 1 in a Wellsian time machine? The choice would be to carefully study and take notes like an anthropologist, or to take maximum advantage of the perspective of a later mindstep. Hernando Cortes went to Central America in 1519 A.D., not as an anthropologist but as a declared colonizer, and nothing was going to stand in his way of getting Mexico for Spain and/or himself. He came from the ordered world of Catholic Europe into an isolated world of mysticism, dark minds and astral powers. Ironically, in the clash of the mindsteps those archaic, zodiacal gods, imaginary though they might be, were to play a sequence of cruel coincidences on the mortals of mindstep 1.

Quetzalcoatl was still worshiped by the Aztecs, even though the graven image of planet Mars, Huitzilopochtli, had somewhat replaced him, and the Quetzalcoatl legend was burned into the Aztec mind. Emperor Montezuma lived by the version that said Quetzalcoatl would return to walk the earth, and his kingdom was gripped by the astral movements in the zodiac.

As it happened, Cortes landed at Tabasco during the long encounter of Venus, the segment of the legend where Quetzalcoatl was searching for his father's bones. Cortes landed 400 warriors, 16 horses, and cannon and gunpowder kegs on the beach. He commanded thunder and lightning to come from the cannon. He stirred up the sand of the beach. Horses had never been seen before in the New World, and Montezuma's forward observers sent back reports to the capital of a man who must be a god because he possessed steel, a metal found only in meteorites, because he commanded thunder and because, like Quetzalcoatl, he was aided by strange beasts.

Cortes took aboard for his wife the beautiful Indian Malinche, who claimed she was a princess, and he hoisted sail and left the coast watched by Montezuma's intelligence network. The Aztecs were relieved to see him go, because for 8 years past there had been ominous signs and portents—3 demoniacal apparitions had moved among the stars and a cone of light in the east reached from horizon to the zenith. The apparitions were probably long period comets, and the cone was the zodiacal light, the glow of sunlight reflected from meteor particles in deep space, but soothsayers predicted, and Montezuma believed, that these portents predicted his downfall.

The first landing during the long encounter might have been pure chance, but now Cortes had the powers of the cosmos if he cared to use them. Malinche knew the details of the Quetzalcoatl legend and hated the Aztecs. She was quick, intelligent, and she fell in love with Cortes. It would be difficult not to believe that she and Cortes used the orbit of Venus to their advantage. She learned Spanish; she acted as interpreter and adviser. She and Cortes were so closely allied that the Aztecs called him "Malinche." As the events unfolded, they should have called him "Cortes, morning star."

The intelligence that Montezuma was receiving back at the capital changed from elation at the departure of Cortes to near terror. The white-plumed vessels went against the north wind up the coast (by tacking into the wind) as if "propelled by serpents." Cortes put ashore at what is now modern Veracruz, unloading horses and cannon. He chose to land on April 21, 1519, by the Old Style calendar, at the beginning of the new Venus cycle (ASTRODATE 1519.303217, Venus AGE 38.6 from computer programs in Appendix.)

Back in the capital (modern-day Mexico City) the priests noted how Venus-Quetzalcoatl had emerged as the evening star. For them, fair-skinned, dust-blackened, bearded god Quetzalcoatl had arrived as a man. Montezuma ordered the Aztecs and the states under his sway to receive Cortes as an emissary of the sky-gods.

The Cortes-Montezuma relationship and the ultimate conquest of Mexico is well documented. We know the details of the

Cortes plans and changes of plans. We know how Montezuma gave secret orders for the destruction of the 400 warriors if a way could be found, and we know almost word for word what Cortes and Montezuma said to each other. The astronomical connection is strong.

Cortes played the good god Quetzalcoatl on his overland journey to Mexico City, peaceful and just, forgiving to his enemies, strong against treachery. The enemies of the Aztecs became his friends, and the Aztecs in the capital marveled at his influence.

After 244 days Venus neared the short encounter, and Cortes arrived in Mexico City. In line with the legend, Montezuma prepared a palace for Cortes and his men, with wine, music and women. But as the Venus-sun conjunction occurred it was Cortes who took the role of "deceitful enemy," and Montezuma who was led to sinning. Cortes invited the emperor to go to live in the guest palace to enjoy the wine and sexual partners, and where he would be a prisoner of the small Spanish army and its allies. Montezuma moaned as he accepted the invitation. "It is ordained!" he said.

By now it was December 1519 and Cortes arranged the public execution of some Aztec nobles on the charge of treason against him. He had them burned at the stake, symbolic of Quetzalcoatl's own death in the fire. While the smoke was rising he went in to Montezuma's room, said to him, "You have sinned!" and had leg-irons put on the emperor to make him do an act of penance. The executions were over. The smoke lingered in the great square, and through the dawn mist Venus showed as the killer morning star.

In May 1520 there was a conjunction of Venus and Mars, and the Aztec priests held a ceremony for Mars-god Huitzilopochtli. Stung to action, thousands of Aztecs besieged the guest palace where emperor Montezuma was held hostage. Cortes ordered Montezuma to intervene, but the emperor said he was weak, had no power, and that he had been forsaken by the gods. Venus was fading in brightness and retreating toward the sun. Cortes ordered him again. Montezuma bathed, put on his finest robes and climbed to the top of the roof. He spoke loudly and his voice could be heard to the ends of the plaza, but the crowd below cried: "He has grown weak, he is like a woman, kill him!"

Suddenly arrows and stones rained down on Montezuma, and before the conquistador bodyguards could raise their shields over him he was struck down. The wound did not seem to be serious, but the Aztec leader did not recover. He lay as if dead for several days, and then, as morning star Venus was fading and going to conjunction with the sun, he died.

Montezuma saw in the Venus-Quetzalcoatl legend his own destiny. He was a priest and a god-emperor of classic tradition, and what was in the collective mind focused with irresistible force from the star patterns onto him. Movements in the zodiac were tragically played out in the mirror of the earth. Ancient Mexico's culture, wealth and treasures of the millennia—an empire of a million priests and warriors—had been conquered by 400 men and a princess. Venus shone down in its cycle and Cortes wielded the sword and political intrigue in synchronism.

It would be wrong to say the pre-Columbians had a simple view of the universe. They did not. The movement of the sun, moon and planets is complicated enough, and they added their own convolutions. The Mayan calendar, used throughout the region, was 260 days in length. I have often wondered why they chose 260. Perhaps, as I figured in a footnote in *Beyond Stonehenge,* they knew that Mars was a brilliant midnight star every 780 days. This might have been the reason for the hitherto unexplained length of the religious calendar—3 sacred years equaling 1 synodic revolution of Mars. Maybe it was an astronomical choice, or maybe there was no rationale behind it save a fascination with the play of numbers. But it was not a good choice for fitting with the seasons, being far too short. The sun slipped relentlessly and it took 73 Mayan years for the natural seasons to return to the starting point in their calendar—73 Mayan years of 260 days precisely equaling 52 years of 365 days. They created for themselves the great 52-year convolution. There were fears of disaster and the world coming to an end every 52 years. Cities were destroyed and rebuilt. The gods required extra-special propitiation at this time. It is impossible for us to understand the panic in their minds. Everything moved in synodic periods—time steps. There was the year of the sun, and the year-step of Mars, and the year-step of Venus. On that special day after 73 sacred years had passed, the sun was 52 steps further on, Mars was 24⅓ steps on and Venus was 32½. For them the coming together of periodicities was a barrier in time, almost impossible for the world to overcome. For the Babylonians it was only an extra line of figures added to the clay tablet, and for us nothing more than pressing the key of a calculator.

Needless to say the calendar had to be watched constantly. The priests were in charge, looking out of the temple windows and along the sightlines. Some of the glyphs show a face peering over crossed sticks or some sort of measuring instrument. The Aztecs set up the round temple of Quetzalcoatl in front of the main pyramid in Mexico City. There were two temples side by side atop the pyramid, and a priest standing at the Quetzalcoatl altar saw the sun pass through the gap on the first day of spring. Father Toribio Motolinia, writing while he was converting the Aztecs to Christianity, said: "The festival called Tlacaxipehualiztli took place when the sun was in the middle of Huicholobos which was at the equinox, and because it [the temple] was a little out of the straight, Montezuma wished to pull it down and set it right." Cortes, of course, interfered with that plan. Both temples were pulled down, but they were never set to right.

We are sure the pre-Columbians knew more than the chroniclers tell us. Toward the end of the Conquest they even gave deliberately wrong information to protect their knowledge.

Astronomers, being a breed of physical scientist, tend not to speculate. The clear road of logic leads ahead and on either side lie swamps and bogs. But there are those who do not see things that way. On either side of the disciplined track are gardens of delight, fountains and meandering pathways. I never know how to answer the question, "Did the Maya have telescopes?" Pre-Mayan mirrors have been found, dull-surfaced to be sure, but the right curvature for a reflecting telescope. The missing piece is the small lens for the eye to look through. Without that we cannot say the artifact is a telescope. When someone points out that the glyph for Venus in the Tajin panels is sectored like the moon, and that Venus through a telescope shows phases like the moon, I am equally baffled. Present-day eyesight cannot detect the crescent shape of Venus morning star—it is far too small. We have no reason to expect pre-Columbian eyesight to be any better than ours. Perhaps, and this is only a speculation, they reasoned that if Venus was really a globe like the moon it ought to show phases like the moon. But speculations require proof, and in this case there is none.

There is an observatory at Chichén Itzá in Yucatán. Called by the Spaniards the Caracol ("shell"), there are windows and stairs which point to the setting of Venus. The grand stairway to the northwest runs directly toward the most northerly point on the horizon reached by this planet. This is one of the stations of the 5-pointed star, and Venus emerged at that spot once every 8 years. The observatory itself is badly ruined, but a few windows remain on the western side. Actually these are quite long; they are more like viewing tubes than windows. A priest looking across the far edge and near edge of the window would see the northernmost setting of Venus in one window and the southernmost setting in the other. We don't know whether there were similar observation lines for the rising planet because the eastern side of Caracol has disappeared.

At Uxmal, another great Mayan city to the west of Chichén Itzá, there are the ruins of what is known as the governors' palace. The building faced southeast, toward the rising point of Venus morning star. In fact, the perpendicular to the wall of the palace lines up precisely with the southern extreme of Venus in the 8-year cycle. The planet would have been seen when looking through the main door. The sight line continues out to the horizon and is conspicuously marked. It passes over a phallic stone (sinning Quetzalcoatl?), a two-headed jaguar altar, and on the far horizon, rising above the jungle treetops, the line touches the ancient, so far unexcavated, temple mound of Nohpat.

Not much is left of the jungle cities. Even less is left of the minds. Some of the temple complexes were abandoned at the end of particular 52-year cycles. Others fell in warfare, and others could no longer be supported by the depleted agriculture. But enough clues are left to show the imprint of the cosmos on those civilizations. Astronomical numbers whirled in their minds, and their lives were dominated by the sun, moon, Mars and Venus-Quetzalcoatl.

FIVE

Stonehenge: A Clue

The ancient Britons and the Maya were linked by one thing—they were in the same mindstep. The Maya were too far away, separated from the next mindstep by a continent and 3,000 miles of ocean. The Britons were too early, separated from the next mindstep by 3,000 years of time. Both of these cultures lived and died under the same celestial vault. Both had imagery, fears and hopes. Both could handle raw stone and build monuments to conquer time. The Maya wrote with glyphs and codices, and their oral folklore was put down in Spanish by the missionaries. But the ancient Britons did not write. Their folklore is almost gone. It was not put down in writing by the Celts, nor the Romans. Today we have only a few scattered clues.

Around 3500 B.C. the proto-Europeans began to build with large stones—megaliths. What impulse drove them to do this we do not know. The stones, sometimes roughly shaped, sometimes in natural form, were set up in rows, in circles or alone on bleak hillsides and valleys as single standing menhirs. Tons of dead weight, a challenge to muscle and mind, seemingly impossible to move, these stones have stood through the centuries as a mystery for the civilizations that followed.

Stonehenge, in southern England, is the eighth wonder of

NOTE: This chapter incorporates new results from field trips to England and Scotland in 1973, 1974, 1980, and 1982.

Stonehenge viewed from the heelstone (right).

Stonehenge from the west, showing the moonset trilithon and the sighting archway.

Stonehenge from the southwest.

the ancient world. In ruins when the Romans conquered Britain, and when King Arthur in fact or legend formed the Round Table at his court, Stonehenge with its air of mystery has inspired poets and storytellers through the ages. Today more than a million people a year visit the site from all over the world to gaze at the mute standing stones, crusted with lichens and worn by the weathering of the years. One can hear almost every modern language there—German, Spanish, French, Japanese, Estonian, Arabic—yet the language of the original builders is unknown. The circle of stone is a quiet inspiration for the visitor, a symbol, a mindstep monument.

Inigo Jones and John Aubrey in the 17th century drew site plans and speculated. William Stukeley in the 18th century wrote about the folklore claim that the sun rose in the Avenue. Sir Norman Lockyer measured that Avenue in the present century to try to date the monument, but his estimate turned out to be a thousand years too late.

My own approach was different. I studied Stonehenge as a whole—postholes, menhirs, archways, rings—and discovered the astronomical pattern. The archways, menhirs and posts lined up with the slow swing of the moon and sun on the horizon. The numbers in the rings matched the timing of these movements.

Unknown to me, a British amateur, C. A. Newham, was working on a limited aspect of the alignments, and I was later able to help him in his work.

I published my theories in 1963 and 1964, and, at least in the first broad brushstrokes, the astronomical side of Stonehenge was decoded. We now had a new perspective on the mystery of those ancient stones.

My interest in Stonehenge had begun in grade school when I read that the sun rose in the Avenue on June 21 and that the Druids had built it. But this was folklore tradition, not hard fact. Archaeologists doubted the Druid connection, showing that the ancient cult could not be traced back further than 500 B.C. (those few "Druids" who worship the sunrise today are a society organized not more than 200 years ago). Archaeologists also denied the sunrise over the heelstone, saying "it does nothing of the sort." Even to think about it, they said, was "fruitless conjecture."

Then I visited the site on Salisbury Plain in 1953 when I was a scientist doing military service with a missile-testing unit 3 miles to the north. At that time, Stonehenge was unfenced and

Inside the sarsen circle.

The heelstone framed by the sarsen archways.

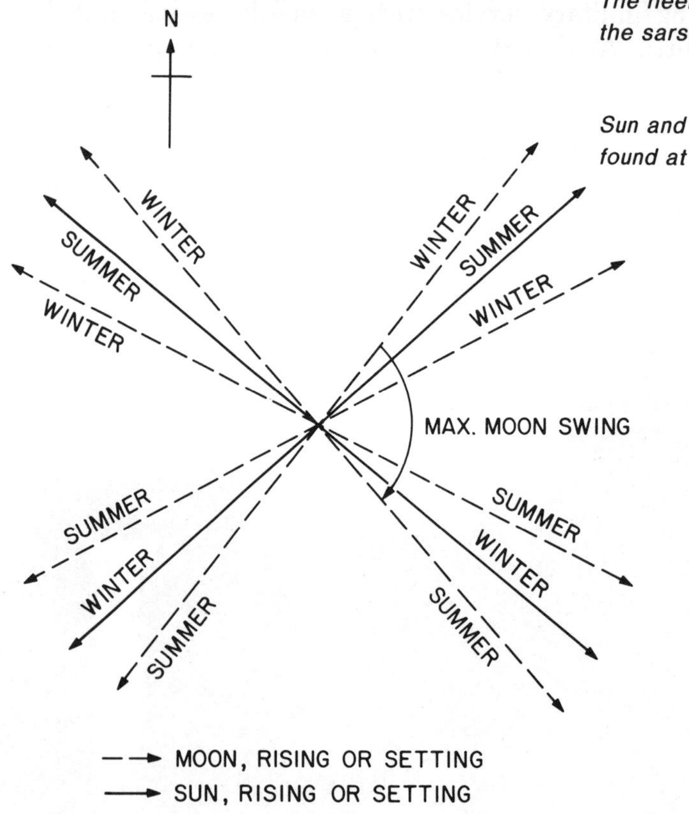

Sun and moon alignments found at Stonehenge.

sheep nibbled at the grass. The winter landscape was blea forbidding, the stones gray and challenging. Why were the ways so narrow—too narrow to walk through—and why did it have a ring of 56 chalk pits around it?

On paper the folklore claim about sunrise looked reasonable enough, but an accurate measurement at the site was needed to check the theory. Once a calibration point was fixed, then other lines could be investigated.

I returned to Stonehenge in the summer of 1961 to test the sunrise with a telephoto camera. As I stood there at dawn I could see through the viewfinder, and later measurements of the cine frames confirmed, that the sun *did* stand on the tip of the heelstone as tradition had it. There was no basis for the doubts. Allowing for the tilt of the stone (now 20 degrees), and the passing years (4,500), the disc of the sun stood precisely on the tip. Armed with this fact I was able to calculate the azimuths (bearing angle from due north) of all the other lines. I found a good match with the rising and setting of the sun at midsummer and midwinter. I also found lines and archway pairs which pointed to the moon. By 1963 I had discovered 24 sun and moon alignments; by 1965, 32. Stonehenge was more than a ritual temple, more than a Stone Age architectural wonder; it contained astronomical and mathematical information, and this knowledge was not supposed to have been discovered until the time of the Greeks a full thousand years later.

It has been claimed that I see the sun and moon in ancient structures because I am an astronomer. Perhaps so. Perception is colored by background and experience; this is unavoidable. I am trained to look for astronomical things which might be missed by others. But I couldn't see these things if they were not there. A modern astronomer has the great advantage of the computer and its ability to calculate how the sky looked thousands of years ago. In the same way the archaeologist is conditioned by his background, and so is the cultural anthropologist. As the disciplines are divided in the United States, the archaeologist concentrates on the artifacts of graves and pottery, menhirs and postholes, and with the help of the cultural anthropologist attempts to fit these into a cultural and social model. In Britain there is a broader field of prehistory which spans both disciplines. Each researcher adds a part of the picture and the parts must be put together to make the whole. But all the experts are faced with the problem of looking back over a vast time span where the only link is the silent standing stones and the knowledge that human beings were involved.

Digging is now forbidden at Stonehenge except for "rescue archaeology." When a new telephone cable was laid along the road from Bath to London in 1979, a companion to the heelstone was discovered by Michael Pitts. When the Department of the Environment made the car park and underpass in 1967, 3 large postholes were unearthed. Before that there were excavations in the 1950s, the 1920s, and at the turn of the century. A lot of important information has been collected, but even so the story is incomplete. The western half of the site has never been excavated. Archaeologists disagree on interpreting the details, and the dates of construction have been revised several times. The broad picture is as follows:

In 2800 B.C. Secondary Neolithic people dug a circular ditch

The telephone trench where the new stone hole was found.

large enough in area to contain Priam's Troy, and built a chalk bank inside with an entrance gap to the northeast. They set up tall wooden posts at the entrance and at other places in and around the site. Some of these posts, judging from the holes, were nearly 3 feet in diameter. The heelstone and other stones probably replaced the wooden posts as the site was developed. There were 56 holes equally spaced in a circle just inside the bank, and extra holes in the entrance (labeled in the illustration A, B, C, D, E) and in the circle (labeled F, G, and H). The 56-hole circle is named the Aubrey circle after the 17th-century antiquarian John Aubrey. Apparently the holes never contained stones but were filled with rammed chalk soon after they were dug. The wooden posts came from the process of forest clearing. The stones were dragged from the Marlborough Downs 20 miles to the north. The work went on for hundreds of years. Later cultures were named the "Beaker People" (because they used a long drinking beaker) and the "Wessex Folk" (because their grave mounds cover an area in southern England once known as Wessex). The Beaker People built the long Avenue, which runs into the site from the northeast. The Wessex Folk built the Stonehenge that we see today, the 5 trilithon archways and the surrounding sarsen ring. They also set up the 19-stone bluestone horseshoe in the center. These stones had originally come from Mt. Prescelly in Wales. The site was finally abandoned around 1400 B.C.

It is natural in mindstep 1 to think of the sky as an enormous dome enclosing the earth. That's the way the Greeks pictured the cosmos—a celestial sphere on which the stars were hung. The sphere was supposed to rotate once per day on perfect bearings, and this explained how the sun rose in the east and set in the west. The stars were fixed on the sphere with the zodiac constellations making a wrap-around belt. The planets, sun and moon crawled slowly on the sphere. Mother Nature provided a giant planetarium.

The sun moves along the center lane of the zodiac, crossing the celestial equator at the spring equinox when it is in the sign of Aries, and at the autumnal equinox in Libra. The sun's track is called the ecliptic. The angle measured up from the equator is called declination and can be figured by pocket calculator (program SUNDEC, see Appendix). On June 21 the sun's declination is

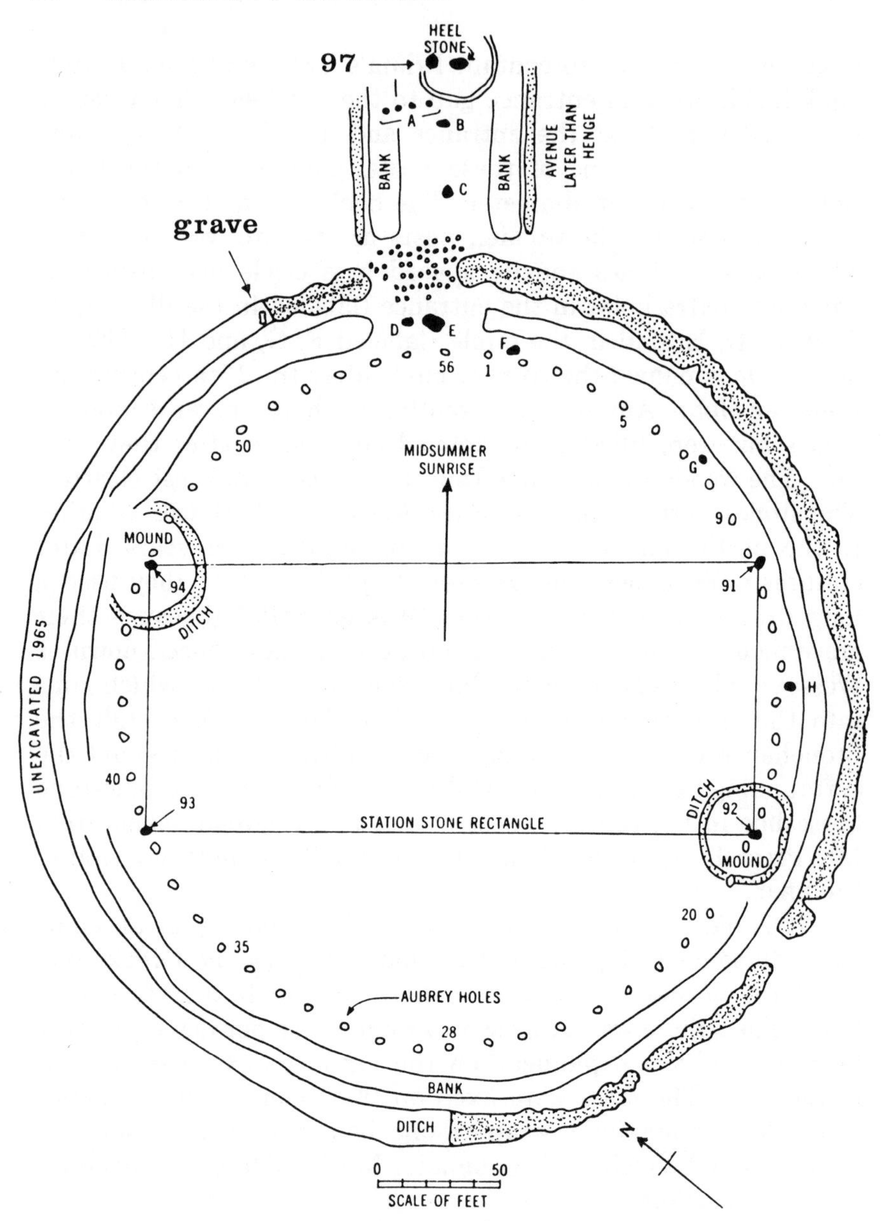

The early "Henge" at Stonehenge, showing positions of newly discovered grave and companion to the heelstone, 97.

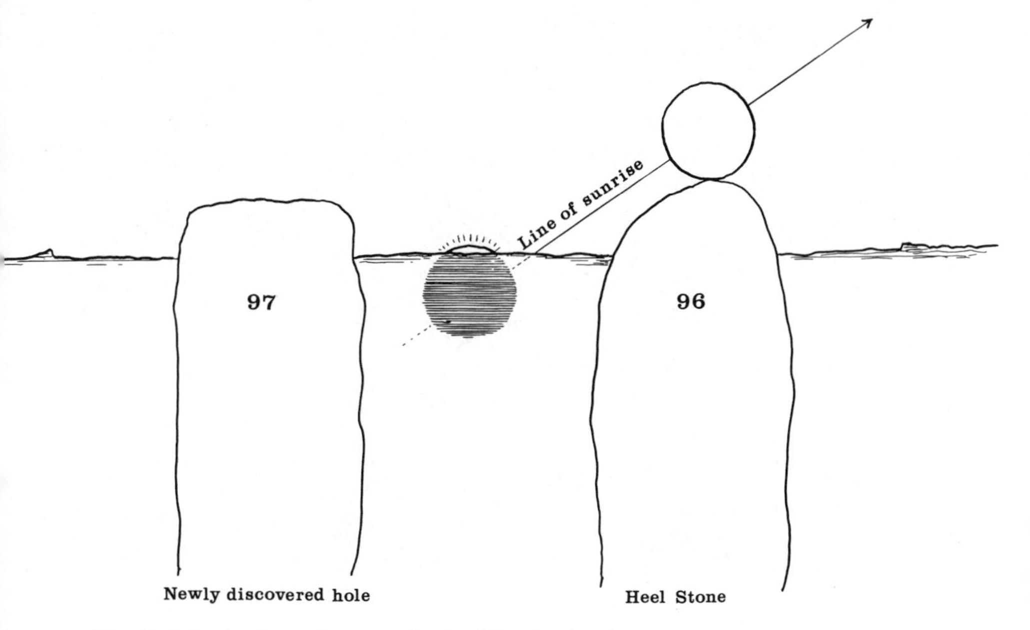

The first flash of sunrise was flanked by the heelstone and its companion as viewed from the center of the Henge.

23½ degrees. At the time of Stonehenge it was close to 24 degrees. In spring it is 0, and in winter −24. This up and down movement is the cause of the seasons. It was considered to be the great disorder of the cosmos by the ancients.

The Stonehengers projected the zodiac onto the flat landscape. They watched how the ecliptic track moved along the horizon with the seasons. They joined the earth to the sky with sight lines and created order from the disorder.

The newly discovered hole near the heelstone shows how the midsummer alignment worked. As June 21 approached, the sunrise position would move northward into the gap between the stones. Because Stonehenge is at a northern latitude, the sun when it rises moves at a slanting angle to the right. On midsummer's day the first flash appeared midway between the stones as seen from the center of the monument, and then the full disc emerged and broke away from the distant skyline. Moments later the disc climbed half a degree moving to the right and stood precisely on the point of the heelstone. Things would look about the same on June 22 and a few days on either side, but neverthe-

A youth sacrificed and buried near the entrance, with his head toward the sunrise, about 2,000 B.C.

less the builders had framed the sun at the solstice and marked the turning point.

I wonder if they felt they had some control over it and the seasons. They lived off the environment, and the sun was vital to their lives. One can imagine the Stone Age shaman looking over the stones, checking to see if the cosmos was in order. There were stones to mark the equinoxes, declination 0, and also the low point in the sun's track at midwinter, declination −24. Dozens of astronomical alignments are locked into the stones and archways of the structure.

Right from the start, the first builders were interested in the moon. To mark the seasonal course of the sun is simple, but the moon is a much more difficult problem. The postholes, now beneath the turf in the Avenue, picked out the yearly standstills of the moon. The sun has 4—midsummer and midwinter risings and settings—but the moon has 8.

When the Stonehengers marked the sunrise with the heelstone and its mate, they found that the sun turned at that spot every year. But the turning point of the moon was not fixed. It moved to the north and then to the south of the heelstone. It swung across the Avenue entrance. So they marked the extreme high and extreme low points with stones D and F, and they marked the years in between with wooden posts. The high-low directions were marked a second and a third time with the station stone rectangle (91, 92, 93 and 94), and the 100-ton trilithon archways.

The swing in azimuth can be figured out with a pocket calculator or home computer. (Use program MOONMAX in the Appendix for declination, and AZIMUTH for the direction of rising or setting. At Stonehenge the horizon altitude is approximately 1 degree all round.) For example, taking the modern era, 1969 A.D. was the year of the high moon (declination 29), and 1979 was the year of the low moon (declination 19). Patience is required to key in the data and get the numerical proof, but even more patience was required by those early observers who stood there, night after night, watching the bone-white disc. Keying B.C. dates into MOONMAX and AZIMUTH will show fairly accurately the phenomena that they saw.

Calculated by eyeball or computer, it takes an interval of 18 or 19 years for the moon to return to an extreme. The period is not an exact, or whole, number of years. Actually it is controlled by the regression of the nodes of the moon's orbit. The nodes regress once around the ecliptic in a period of 18.614 years. This 18/19-year indeterminacy is not gotten rid of until 3 complete swings across the Avenue entrance have taken place. Then the cycle locks into an exact number, 56. This repetition holds regular for centuries. On the average the cycle goes 18, 19, 19, with 2 intervals of 19 years for every 1 of 18.

For example, MOONMAX gives a high moon (29) in the years 1969, 2025, 2081 A.D., etc. It gives a low moon (19) in 1923, 1979, 2035. Every third swing is separated by 56 years. The intermediate swings also have a 56-year pattern: high, 1988, 2044, 2100; low, 1997, 2053, 2109. More patient key-punching will show that after about 6 intervals of 56 (3 centuries plus), there is a 55, and the cycle is reestablished. For people watching the swing of the moon and counting off the years, 56 would be an important number.

Modern-day astronomers do not go out to observe the moonrise year by year. The 56-year pattern, based on the 18.6-year moon cycle, was not known to them until it was uncovered at

Stonehenge. But the cycle is an important one humanistically. During years of the high moon, every 18 or 19 years according to the cycle, the winter nights are brighter and the moonlight lasts longer. In the summer the full moon skims low on the southern horizon. In northern Scotland the winter moon would never set, and the summer moon would never rise. Because the moon affects the oceans, extreme low tides recur after 56 years. If an ancient culture looked on the moon as important either in agriculture or hunting, fertility or death rites, the 18.6-year cycle and the 56-year pattern would ultimately be discovered.

In my opinion this is the reason for the 56 Aubrey holes. There is one for each year of the cycle. The numbers in the other circles at Stonehenge are also connected with the moon. There were 30 archways in the sarsen circle. Since one can't build half an archway, 30 is the best fit to the lunar month of 29.53 days. There are 30 Y-holes and 29 Z-holes below the turf inside the Aubrey ring. Using them together, counting one set and then the other, one gets a month averaging 29.5 days. The 19 stones in the bluestone horseshoe count off the most frequent interval in a single moon swing. The circles have all the qualities of a counting device for keeping tabs on the moon. Whether it was ever used that way we shall never know. Several experts in anthropology have doubted that Secondary Neolithic people could have reached such a stage of development. But the suggestion is still valid from the point of view of the astronomer. And it is a hard-fact explanation for the otherwise mysterious activities at the site beginning in 2800 B.C. and continuing for over a thousand years.

The theory that the Stonehengers were aware of the sky, were watching the sun and moon, were marking the movements and had figured out the periodicities, has been hotly debated ever since it was first put forward. One hang-up in the argument comes from the very basic differences between physical and sociological science—The quantitative and the qualitative. Among some archaeologists there has been the notion that the theory could be proved—QED. "Why! since this is all astronomy and numbers," they said, "let us wait until the theory is established!"

There was the thought that by some complicated theorem, in probability or something similar, the proposition could be proved mathematically. But this is not possible. The theory is indeed quantitative astronomy, but it rests on foundations of qualitative sociology. Probability equations apply only to random structures, and Stonehenge was not set up in a random fashion. No, the Stonehenge results cannot be "proved" in the sense of a geometrical theorem. But then, neither can they be "disproved." As a theory about the mind of ancient Britons it belongs to the fields of the humanities, and as such it requires being generally accepted, rather than proved. Here one has to draw on many disciplines—cultural anthropology, archaeology, mythology, sociology, folklore. Does the new theory fit in with the general picture? In the 1960s the theory was almost iconoclastic, but later discoveries and a long and careful debate have come down on the side of the ancient Britons. They and the proto-Europeans were indeed smarter than we had imagined. They had reached mindstep 1.

Part of a test of a theory is what is discovered afterward. New evidence is always critical in the debate. The 3 postholes uncovered when the car park was built added some pluses. If they originally held tall posts, the 3 holes would have marked the moon as seen from station stone 93. In fact, these holes are very closely on the line of the edge of the rectangle, 92–93. The first posthole marked the moonrise's first gleam, the second marked the center of the disc, and the third marked the moon's orb standing tangent on the skyline. That gave 3 pluses. On the minus side, radiocarbon dating of scraps of bark at the bottom of the holes gave a date several thousand years before the heelstone. Could work at the site with crude posts have begun so early? Perhaps so. One has to remember that the 56-year pattern seems to have been known at the beginning phase of the "henge", and prior observations would seem a reasonable thing to find.

In 1972 Richard Brinckerhoff of the Phillips Exeter Academy asked me if archaeologists had ever looked for carvings on the top of the lintels. I said, "To the best of my knowledge, no." His question was well put; somebody should have looked on top. And so Brinckerhoff did. The British Department of the Environment graciously gave permission, provided no more than one sneaker-shod researcher stood on top at any one time. The custodians provided scaling ladders and up he went. The trilithons were blank, but the sarsen lintels over the northeast archways were not. There were 11 pits in the surface. The first pit over the left-hand archway as viewed from the center was 6 inches in diameter and scooped out 3 inches deep. Fortunately for the theory, the pit marked the extreme northerly moonrise in the 18.6-year cycle when viewed from the great trilithon. The observer either could stand or sit on top of the trilithon, or could stand on the top of the sarsen ring behind and look through the gap. I would suppose a stone with a plug-in base was placed in the rounded pit to mark the moonrise. This would not have looked very neat architecturally, but then the theory says Stonehenge is more than an architectural marvel anyway. Brinckerhoff wondered if the Stonehengers used the circular track on top of the sarsen ring as a walkway, an astronomical platform. Perhaps so, but it is difficult to say. The whole southern sector has disappeared, ravaged by time and plunderers.

When the moon is in front of the sun, the sun is blotted out. When the moon is opposite the sun, the moon is blotted out.

Marker holes discovered on top of the lintels.

Eclipses do not happen very frequently—only at 2 seasons of the year when the sun is at one or the other of the nodes. If the nodes are in the signs of Aries and Libra, eclipses take place in spring and fall. If in Cancer and Capricorn, the eclipses are in summer and winter. For the ancients a dragon lived at each node waiting to swallow the sun or moon. Gilgamesh and Enkidu would meet with the dread monster Humbaba.

It would be wrong for an astronomer not to point out a further possibility at Stonehenge—eclipse prediction. Everything was there to do this. When the winter full moon rose over the heelstone every 9 or 10 years, it was in danger of an eclipse. When it was over D or F, every 18 or 19 years, eclipses occurred at the equinoxes. There is a 56-year pattern for the eclipse seasons. If eclipses of the sun and moon occur in March (spring) 1969, they will occur again in March 56 years later. This is not a familiar pattern to modern-day astronomers because it predicts only the season of eclipses, not a precise, actual eclipse. But these seasonal danger periods would be a useful thing to know about in Stone Age times, if there was concern for the light of the moon and a sudden "going out."

It was in the 1960s that I published the article theorizing that Stonehenge might have been a "Neolithic Computer." The idea was too early, then, for acceptance. It was a time when computers were generally suspect, something like the dread Humbaba. In my article I was obliged to acknowledge the donation of one minute of computer time from the Smithsonian Astrophysical Observatory where I did the work, and that identified me, for some readers, with Humbaba.

My critics jumped on the noun "computer" in the title, and ignored the adjective "Neolithic." I had nothing more in mind than a basic counting device built by Stone Age people, but the computer notion stuck. Since the 1960s there has been a change. Electronic digital computers are better understood for what they are—calculators, fast and accurate, processors and storers of information, creators of art forms, game-playing gadgets, etc. As it was in the 1960s, the Smithsonian machine, though it could compile and compute dozens of alignments in less than a minute, was managed by what was called "batch processing," and a person had to wait a full 24 hours for the output to be returned. That basic program for astro-archaeological alignment is given in the

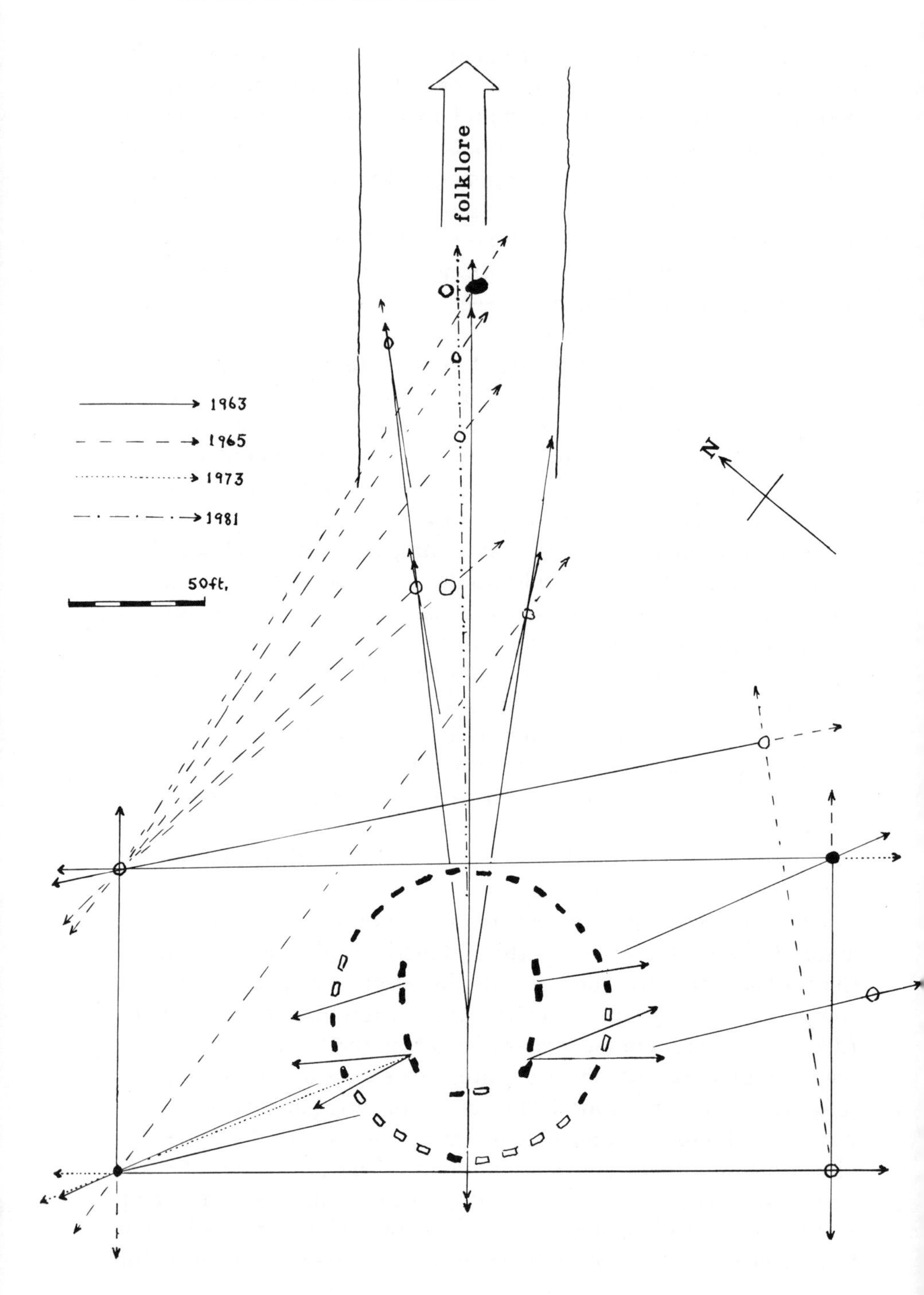
folklore
1963
1965
1973
1981
50ft.
N

Appendix (STONEHENGE). It fits into 150 steps of a progammable wafer. Nowadays it takes 15 seconds to get an answer on a pocket calculator as the lights in the display window go out and the silicon chips go to work. Future models will undoubtedly be faster.

Yes, Stonehenge with its numbered circles and sight lines was an ideal device to warn of "danger" years. With it a Stonehenger would have the power to say in which month the moon was likely to be eclipsed. Every 9 or 10 years this would happen in the month of the winter moon. The same system would work for eclipses in spring, summer and fall, according to the overall 18/19, 56-year pattern. I spoke about this at a high school in Virginia several years ago, when someone there made a very interesting suggestion. "Maybe it worked the other way around. Maybe they wanted to know if the moon was going to be *free* from eclipses. This way they would be right for 8 years in a row and would only have to get worried on the 9th." That made sense to me. It was a different perception. Like looking at the doughnut instead of at the hole.

Maybe it's a computer, said the critics, but was it ever used? If Stonehenge were isolated in time and space we could dismiss it all as a fluke. It could be just chance that all the numbers and lines fitted the astronomy. But there are other stone circles. Not rings with shaped lintels on top—that feature is unique to Stonehenge—but there are other megalithic sites which show evidence of an interest in the sun and moon—the astronomical connection. There was some driving force that made the people of the late Stone Age move enormous boulders and set them toward the sunrise. They built hundreds of circles in England, Scotland and Ireland. It was hard work. The Stonehengers put close to 2 million man-days of work into the structures. According to legend it was done by magic, and wizard Merlin was involved....

★ ☆ ★

"Merlin said to the king of England: 'Send for the Dance of the Giants that is in Killaraus, on a mountain in Ireland. A structure of stones is there that no man of this age could lift save his wit were strong enough to carry his art. The stones are big and full of virtue, and if they be set up here in a circle as they are in Killaraus, here shall they stand for ever.'

"The king laughed and said: 'How can stones of such bigness and in a country so far away be brought here, as if Britain were lacking in stones enough for the task?'

"Merlin replied: 'Laugh not so lightly. In these stones is a mystery and a healing virtue against many ailments. Giants brought them from the farthest ends of Africa and set them up in Ireland. Not a stone is there that lacketh in witchcraft.'

"Fifteen thousand men tried to move the stones with huge hawsers, ropes and scaling ladders. But they could not move them. Never a whit the forwarder. Merlin stood there and burst out laughing. He put together his own engines. He toppled the stones over so lightly that none could believe their eyes. Then he ordered the men to carry the stones to the ships to take them to England.

"When the men arrived they were joyful and they set the stones up around the compass of the burial ground just as the stones had stood on Mount Killaraus. This proves once again how skill and magic can overcome strength."*

★ ☆ ★

The legend has an echo of truth in it. There *are* stone circles in Ireland, more than a hundred of them, and they *are* connected with dancing. One circle in County Wicklow on top of a hill is called the "Pipers Stones." According to local folklore, a group of dancers and the piper were turned to stone. There is another piper's stone ring in Kildare (Killaraus?). When some of the Irish circles were excavated archaeologists found a cobbled pavement suitable for ceremonies, including dancing. Then again some stone rings are called "maidens" and "ladies," and when there is an isolated stone outside the ring it is sometimes called the "piper" or the "devil."

Following up the Stonehenge clue, John Barber measured 30 stone circles in County Cork and County Kerry. The circles were small, about 25 feet in diameter, and each contained 12 stones or less. The stones were graded in height, the largest at the northeast, smallest at the southwest. The southwest sides of the rings were closed off with a large horizontal stone, the recumbent.

* Geoffry of Monmouth, *Histories of the Kings of Britain,* trans. S. Evans (London: Dent, 1904).

Usually this closing-off stone was a noteworthy specimen either because of its (what we call) geological nature, or because of its shape.

When Barber measured the skyline in the direction of the recumbent stone and calculated the astronomical declination for the construction date (2000–1500 B.C.), he found parts of the Stonehenge pattern—the lowest sunset (−24) and the 18.6-year moonset (−29). Not all the circles showed the sun-moon connection. Thirteen out of the 30 had no identifiable target in the sky. One or two of the others were set too far to the south to point to the 18.6-year moonset, and Barber wondered if those might not point to the planet Venus. This would be a reasonable extension of the Stonehenge pattern, since Venus is the next brightest object in the sky after the sun and moon. But the direction of pointing from the center of a small circle over the long recumbent is difficult to fix and we need more evidence of Venus alignments to be sure. The Venus extreme shows up once every 8 years on the 5-pointed star pattern in the zodiac, as marked by the pre-Columbians at the Caracol and at Uxmal.

From those circles that marked the sun at midwinter, the sun would be visible every year on December 21 by our calendar, setting over the recumbent stone. For the circles that marked the moon, it would show up only once every 18 or 19 years at the extreme low point of the cycle, full in summer or new crescent in winter. Even though the circles in Ireland are not set out as accurately as Stonehenge, they could serve the purpose of fixing the season for a ceremony. There is evidence of ceremonies. Most of the circles have burials in the center—cremations. Archaeologists found hard-packed soil around the circles—evidence of dancing or heavy treading—and charcoal and burned bones. A large funeral pyre had burned on that dark evening some 4,000 years ago as the weak sun needed rejuvenation, or as the moon skimmed along the mountain tops, coming very near the earth and signalling the time for magic and witchcraft.

Perhaps the ideas flicker in the mind today. The Gaelic farmers are known for their isolation and independence. One does not have to go far down the winding road to find a person who believes in leprechauns and the little people who live inside the hills. When they saw what John Barber was doing in the circles they said: "So you're interested in the Duibrey, eh?"

"What's the Duibrey?" asked Barber.

"Why, that's every nineteen years when the moon stays below the mountains."

One of the Irish circles is large and about 500 years older than Stonehenge, according to radiocarbon dating of cremated bones. The site is New Grange, 30 miles north of Dublin on a pleasant hill overlooking the green Boyne valley. The name is a modern corruption of the old Gaelic *An Uamh Greine,* meaning "cave of the sun."

The stones in the circle are rough boulders weighing 30 tons apiece. Most of the rock is local, of the type found within a 10-mile radius, but some of the stones were dragged from 50 miles away. The word "cave" probably refers to the burial mound in the center of the ring. The mound was very carefully made. The layers of soil, rock and sandy silt were chosen for quality and packed down so well that the inside of the mound is watertight—no small achievement for the present-day climate. The floor of the tomb is always dry. Light dust rises, as in the center of the Giza pyramid in Egypt. The builders lined the sides of the passage into the tomb with upright slabs, and the roof was corbeled over. Corbeling is a bricklayer's or stonemason's way of closing a roof. Each stone layer is displaced sideways until the gap at the top is small enough to be closed by a capstone. This was another engineering "first" for the Stone Age builders. The date was 3300 B.C.

The builders flanked the 70-foot passage with small burial chambers. When it was excavated in the 1950s, archaeologists found stone dishes with human remains that had been placed in the tomb 5,000 years before. The passage was slightly curved and the floor sloped upward toward 3 spirals carved in a chamber at the end. There was a sun symbol, a circle with rays, carved on a flat curbstone at the entrance.

But New Grange was more than a tomb. It was never completely sealed off. The builders left a light box over the entrance, a rectangular window 5 feet wide and 2 feet high. Even though the passage was slightly curved and the floor sloped upward, they arranged the light box precisely to let the sun's rays penetrate to the end chamber on midwinter's morning. Why did they do this? How was death connected with the sun? What did the symbolism mean to them? It was a widespread culture. There are

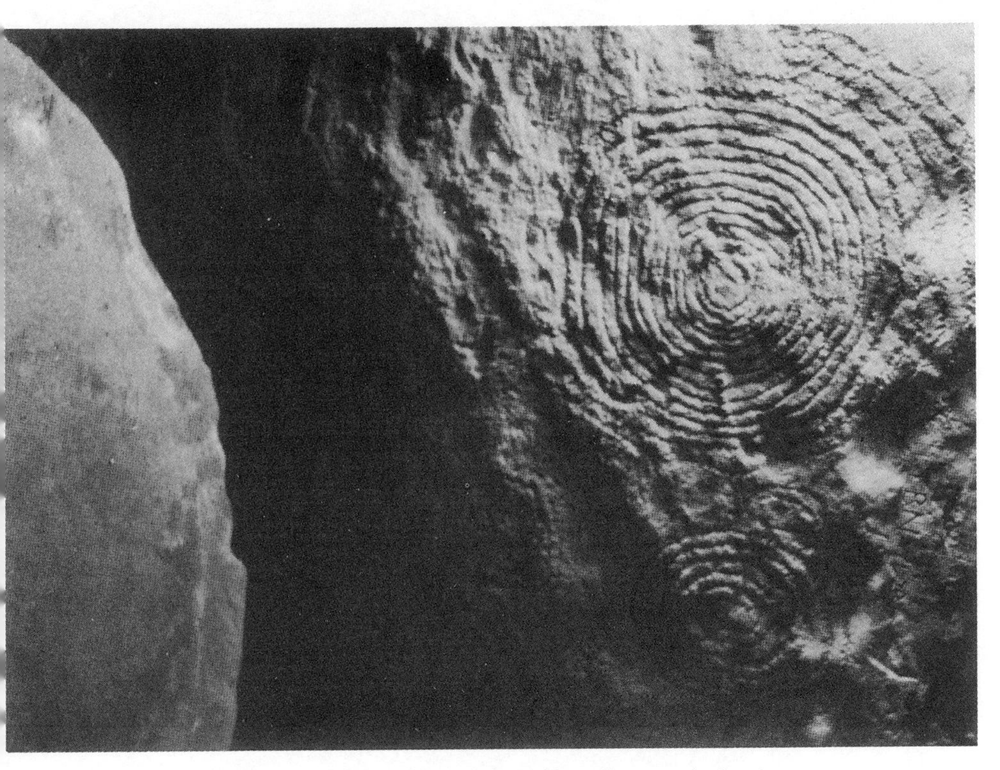

Spirals cut in the end chamber, New Grange, Ireland.

sun symbols on the curbing of a still unexcavated mound at Dowth near New Grange, which is almost certainly a passage grave with a sun orientation. The culture probably went through rituals at the winter season, maybe to ensure the rebirth of the sun when the disc reached the lowest point in the zodiac. Did they believe the dead could be raised in the same way, or that spirits were there at the end of the ray inside the man-made hill?

Aubrey Burl was the first British archaeologist-prehistorian to accept the pattern of sun-moon alignments as valid. So far as he was concerned I had "made an important step forward" with the Stonehenge work, and he based his opinion on his broad knowledge of British stone circles. There are 42 well-preserved recumbent stone circles in Scotland, most of them in Aberdeenshire, and Burl has examined them for astronomical clues. They are similar to the Irish circles of Cork and Kerry. However, the Scottish circles are larger (averaging 50 feet in diameter), the recumbent stone is bigger (up to 50 tons) and the date of con-

Sun symbol on the curb of the passage grave at Dowth, near New Grange, Ireland.

Midwinter sunrise through the roof box, New Grange.

struction about 500 years earlier than the Irish ones. The recumbent was placed in the southern sector and the rings usually contained 12 stones. In Ireland the stones were graded in size with the smaller stones placed next to the recumbent. In Scotland the ring builders had it the other way round. The stones are bigger as one moves toward the recumbent.

According to Burl, every circle in the group has an astronomical connection. The critical line is from the center looking out over the middle of the recumbent stone. One of the circles marks the midwinter sunset, and 7 fit a moonset. The rest of the circles, 34 of them, seem to fit a new type of alignment. They point to the arc of the midsummer moon. Aberdeenshire is at a high northern latitude (57.5 N) and the moon can seem to be very close to the horizon under extreme conditions. These occur once every 18 or 19 years when the moon is in the extreme south. Even with a flat sea horizon, the moon gets no higher than 3½ degrees, and with the ubiquitous distant hills and mountains of Scotland it would literally skim along the skyline. The arc of the moon is low and the sector between moonrise and moonset shrinks to about 20 degrees on either side of south. The 34 recumbents cluster in that sector. Burl believes the circles were made to point toward any part of the low arc, so long as the moon skimmed across the view of the recumbent. But detailed surveys at the sites might show greater discrimination on the part of the builders. When the height of the hills is taken into account, many of the circles might well prove to align on the place where the skimming moon touched the skyline.

There was a large amount of silica in the circles. This glaring white rock must have been collected and carried to the sites. It was scattered near the recumbent stones as if to represent the whiteness of the moon and/or the bones of the dead. There were shallow circular holes drilled into some of the stones—cup marks. These marks were mostly on the recumbent stones and the stones on either side, and so the cup mark in the Scottish circles might be a symbol for the moon. Burl couldn't find any evidence for planet Venus, but it is just possible that cluster of 8 circles near azimuth 195 degrees might turn out to be directed toward Venus.

What can one say of this Irish-Scottish connection? Prehistorians give great weight to similarities. When the same features occur in two different places in a specific geographic area the cultures are surely connected. There is a similarity in the pottery, and the circle design is almost the same. The large recumbent stone is a telltale feature, and both designs point to the sun and moon. The two cultures cremated bodies and held ceremonies within the rings. There *must* have been a connection. Yet the only route is by sea and the passage is not easy, nor convenient. The tides to the west of Scotland are strong and dangerous, and the other route through the English Channel is long and treacherous. Whichever route was chosen, prehistorians believe the migration took place over the centuries and was caused by overpopulation and depletion of the arable land. Another reason for the migration was a search for deposits of copper and gold. The trend seemed to be from Scotland to Ireland. In Cork and Kerry the circles are near ocean inlets, and as one goes inland the circles are fewer and smaller. I am sure the groups knew nothing about modern navigation, yet ironically the great circle track between the old home and the new was the true azimuth of the midwinter sunset. Whenever a group landed—and sometimes, judging by the spacing of the circles and the available arable land, the group was no more than 30 strong—it consecrated the new ground with a magic circle pointing to the setting moon or sun, and conducted their occult nighttime ceremonies around a funeral pyre.

The story of the circles is not yet complete. The altitude of the skyline must be measured at each site, calculations made. This is a time-consuming task, but without those measurements we will not know for sure the level of deliberation in the choosing and laying out of a site. We need more digging to locate missing stones and to retrieve artifacts. The clues are sometimes located outside the circle. For example there are two outlier stones at the circle on the Loanhead of Daviot in the Grampian hills of Aberdeenshire. One accurately marks the midwinter sunrise as seen from the center, the other the equinox. The longer the line, the greater the accuracy one can expect.

There are circles all over Scotland—in the heather by the lochs, on spurs of hills overlooking the ocean, in the Highlands and in the sheltered, loamy valleys. At the last count the number was 407. There are also rows of stones and isolated menhirs. Some monuments have been dug out of the peat beds, others may be buried yet. This brown, soggy, compacted vegetation is dug out by the crofters for fuel. It grows at a rate of between 1 and 2 feet per 1000 years, and a 6 foot menhir can disappear from view and be forgotten until the peat-digger's spade strikes the top of the stone. Thanks to the peat and the pristine isolation of the sites, there are more surviving Neolithic and Bronze Age monuments in Scotland than in any other part of the British Isles.

In 1964 I received a letter from a retired professor of civil engineering, Alexander Thom, who lived in Argyll and who had spend almost all of his spare time surveying and measuring the monuments. He had found alignments with the sun at solstice and equinox, and with the rising and setting of the stars. He asked me if I could suggest a star for those sites which pointed to declinations 28 and −30. My immediate suggestion was not a star but the moon at the extreme of the 18.6-year cycle. The alignments, I believed, were to the moon at the extremes, +29 and −29. (Thom later called them "standstills.")

Because of parallax, the moon's disc is displaced downward on the celestial sphere, and when it is at declination 29 seen from the center of the earth, it seems to be at 28 for someone standing on the surface. In calculations one has to allow for this. Also the moon's orbit is tilted like a phonograph record badly placed on the spindle. In the years of high moon, the winter full moon in the tilted orbit reaches the farthest north position, +29. But in the same year the summer full moon is farthest south, −29. These various tilts and changes produce the Stonehenge pattern. It shows up most clearly in the rectangle of the station stones. The summer full moon rises on the long side of the rectangle in the year of the high moon. Six months later the winter full moon sets on the same line.

I have always been suspicious of star alignments because of practical difficulties. An average star can be seen rising and setting only when the skyline is fairly high, as with a distant mountain range. There are hundreds of possible targets and a high chance of finding a star on a line of stones purely by accident. Also stars are not very impressive objects when they are rising or setting. Even Sirius, the brightest star, fades and flickers into invisibility before it reaches a flat skyline like a sea horizon. And one needs to know the date of construction of the monument accurate to 100 years, because a star position can change by 1 degree per century. This is caused by the earth's precession, the conical, top-like gyration of the spinning axis. The moon and sun on the other hand are not affected by the conical precession, only by the slow change in the angle of tilt, and the exact date of the monument is not critical. Stonehenge, for example, still functions today as a sun and moon marker, and it was built 4,000 years ago.

One of the Scottish sites is at Callanish on the Island of Lewis in the outer Hebrides. A 13-stone circle sits at the end of a low, sloping, windswept ridge overlooking Loch Roag. A stone-flanked avenue approaches the circle from the northeast, and 3 rows of stones roughly east, south and west complete the pattern of a distorted cross. All this was uncovered by peat diggers in the 18th and 19th centuries. Thom thought the avenue was built to point to the northern rising of the star Capella. I suggested that the alignment really worked in the opposite direction, toward the southwest. The computer gave a moon declination of −29, the southerly extreme. The other arms of the cross marked the equinox sun and the middle of the moon's low arc. Thom graciously agreed with the change, and proposed further that the builders intended to use Mt. Clisham as a distant foresight for precise observation of the lunar disc, accurate to minutes of arc at the very limit of resolution of the human eye.

Callanish is near the arctic circle of the moon. Every 18th or 19th year the full moon in summer barely rises, skimming bird-like along the distant mountain tops. If Callanish were farther north, the moon would be below the horizon, invisible—a remarkable place for observing the moon. One wonders if the builders chose the site because of its arctic-moon latitude, or whether, happening to live in the area, they marked the strange event, unaware of what would be seen somewhere else on earth. Whatever the motive, the astronomical facts stand out as clearly as the stones: the distorted cross on the ground matches the sky for this unusual latitude.

When I visited the site and made exact measurements I noticed a difficulty with the mountain foresight theory. A small rocky outcrop prevented a clear view of 20-mile distant Clisham from the center of the avenue. The mountain could be seen from the end of the ridge and also from the right flank of the avenue, but not from the center. This might present some difficulty in Thom's theory about the mountain's use as a foresight, but not if the observer walked back down the avenue a few hundred feet. From there the top of the mountain is just visible over the low outcrop. Fortunately my own suggestion was unaffected. I felt the avenue itself was the marker, and the summer moon set on the line of the stones whether the tip of Mt. Clisham showed or not. I walked back to the cottage where I was staying, careful to avoid the deep peat trenches filled with red-brown water.

One attraction in the field of astro-archaeology is the travel. A researcher applying astronomy to the study of ancient structures must go to the site, wherever it happens to be. There is no substitute for on-the-spot measurements, no matter how accurate the survey plans are. But of course, travel is expensive and the detailed research is time-consuming. The National Geographic Society supported my work at Stonehenge and Callanish. At Kilmartin, Argyllshire, I went with a team from Educational Expeditions International (EEI), now known as Earthwatch. EEI provided jeeps, theodolites, and a radar-operated tape measure. The hard-working team was made up of volunteers—teachers, students, psychologists, and a Philadelphia bank president.

The hamlet of Kilmartin is 10 miles inland from the noisy Corryvreckan whirlpool at the head of the Sound of Jura. The Kilmartin stones are thick gray slabs set in a row with a pair at each end and a 15-ton menhir in the middle. There is an isolated stone to the northwest, and under the trees of Temple Wood a circle of the same gray slabs. The stones are set on the flat floor of a cultivated valley. When EEI arrived, the long fields had barley and rye, oats and hay, but the landowner harvested the hay earlier than normal to help the team explore. We were able to work in the other fields without disturbing the crops by using the radar measuring tape.

If Aubrey Burl was right and cup marks stood for the moon, the Neolithic builders had certainly left an obvious clue. The central menhir was covered with cup marks beaten into the hard rock surface with a Stone Age chisel. But there were not many archaeological facts known. The date was approximately 2000 B.C., and the Temple Wood circle had been paved with cobbles and contained some cremated burials. Originally there might have been an extra menhir between the one with cup marks and

Temple Wood circle, Kilmartin, Scotland.

the pair to the south. There were 3 stones which might have surrounded this menhir at one time. We called these stones the "small ring."

The surveyor on the team, Jon Patrick, and the assistants determined due north by sighting the pole star, and measured the XY coordinates of the stones. Since the values have not been previously published, I give them here. The X-axis was surveyed to be exactly east-west. Units are in meters, height above sea level 30, latitude, 56.1 N.

Position	*X*	*Y*
center of north pair	1023.55	1064.19
cup-marked menhir	1009.79	1032.12
center of small ring	1006.88	1023.64
center of south pair	995.34	999.22
northwest menhir	967.20	1159.63
center of Temple Wood circle	800.85	1240.28
newly discovered stump	699.41	1007.38

When the measurements were completed and the calculations done, we found that the Kilmartin stones pointed to the moon, −29. In fact the Kilmartin line pointed the same way as the Callanish avenue. They both marked the lowest summer moonset. The mainland and the Hebrides Island structures were parallel astronomically. As with Callanish, other stones at Kilmartin made other alignments. The line from the northwest menhir to the cup-marked stone fitted the lowest summer moonrise. So once every 18 or 19 years the full moon rose over the cup-marked stone as seen from the northwest menhir, skimmed along the horizon and set over it when viewed from the northern pair. Whether or not the cup marks were symbols for the moon, the stone marked the end-points of the moon's summer arc. And sighting through the pair of stones at the end was reminiscent of the heelstone and its twin at Stonehenge with the midsummer sunrise.

Cup and ring marks carved by prehistoric Britons on an exposed slab near Kilmartin.

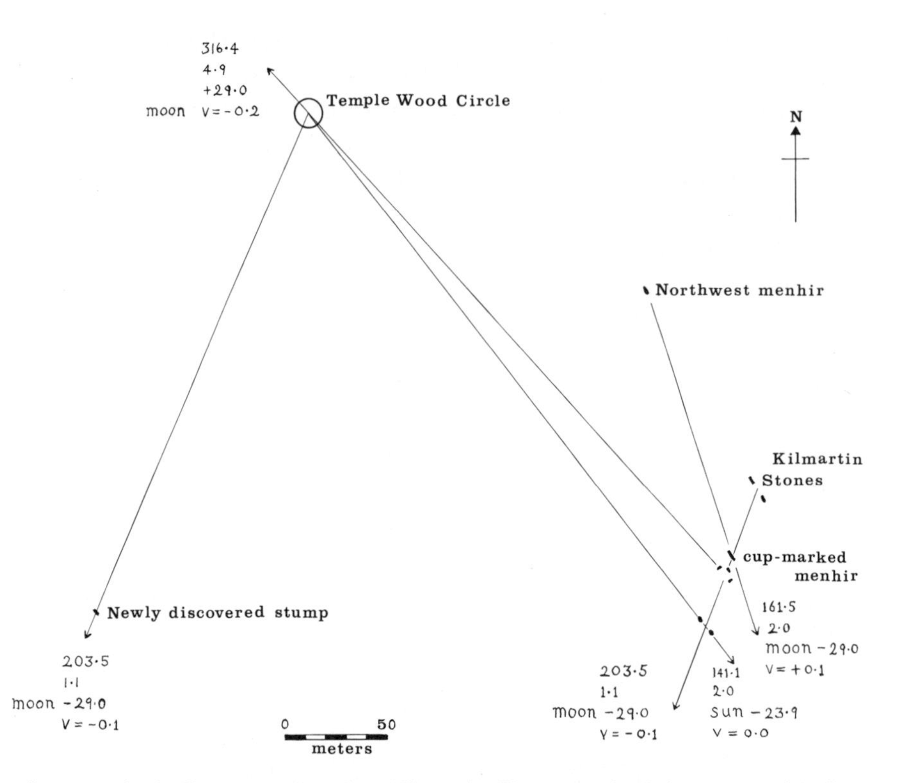

Astronomical alignments found at Kilmartin. The numerical data are azimuth, skyline altitude, target declination and vertical error for the sun or moon standing tangent on the skyline. (Program STONEHENGE, *latitude 56.12, height above sea level 50 meters, assumed date 1,800* B.C.*)*

Today it is not easy to get from Glasgow to Lewis. British Rail connects by boat to Skye and by intermittent island ferry to the island of Harris, which joins Lewis. It was less easy for Neolithic peoples, and they had to worry about the Corryvreckan whirlpool in their small, wood-framed, skin-covered boats. Yet here was a connection in the stones, a link between the minds of those people.

Archaeologists are wary of mixing different cultures, different monuments. In the Kilmartin valley there is no way of telling whether the Temple Wood circle is related to the row of stones. But astronomically there seems to be a relationship. As seen from the small ring, the winter high moon, +29, set over the center of the Temple Wood circle in 2000 B.C. Actually the viewing point *might* have been marked by a tall menhir—we can't be sure—*and* it might have been cup-marked! Also, seen from Temple Wood, the midwinter sunrise came up over the southerly pair of the Kilmartin stones.

Astronomically the alignments were built as accurately as we could measure, given the rough unworked nature of the upright slabs. Archaeologically the stones selected in the making of the Temple Wood circle were the same gray rock used in the row. This was suggestive; maybe the structures were linked, but we needed more evidence, fresh evidence.

When the hayfield was mowed the tractor operator reported a rock. It wasn't new to him. That rock had broken his plowshare and snagged the disc harrow in previous years. Aubrey Burl was with the team and decided to excavate. He made a neat 1-meter-square incision in the stubble with sides arranged north, south, east and west. Volunteers on the team assisted him. It was slow work lifting, sifting the soil. The bank president enjoyed it most, saying it was "one job where you start at the top and work your way down to the bottom."

The plow-breaking rock turned out to be a large prehistoric menhir stump. The tall standing stone must have broken off at ground level centuries before, but the base was still in position. We could see the packing stones at the bottom of the hole, which were used to prop the menhir upright when it was first erected. It was the same type of gray stone chosen for Temple Wood and Kilmartin.

Archaeologically the stump was related to the Kilmartin stones. Even though it was 300 yards away there was a telltale clue—the direction of the slab. The cup-marked menhir and all the other stones were set to point to a bearing angle of 329 degrees. It was a deliberate choice on the part of the builders, an architectural feature. The line through the north pair and south pair have this bearing, and it is the direction of the line to the northwest menhir from the north pair. The stump fitted the original design. The slab was set to the same bearing angle.

Astronomically the stump was an acid test for the theory. If the new stone did not fit the astronomical pattern, then perhaps the other alignments were nothing more than chance.

The coordinates were determined, the calculations made. The new stone linked up with the summer moonset as seen from the circle, and for good measure there was a peak, Ballanoch

Stump of a moon-marking stone discovered at Kilmartin, Scotland.

Hill, directly behind the stone. The new stone confirmed that sightings were made from the center of the circle and that the Kilmartin row and the circle had at some phase of prehistory been used for the same purpose. Neolithic people in their circle ceremonies would have seen the full moon balanced on the peak of the hill at the southerly extreme of the 18.6-year cycle. The outlying stone was a record, a reminder for them. Burl was delighted with the stump and packing stones; I was pleased with the astronomy. Surely now there could be no doubt that the moon was dramatis persona in the minds of the ancient Caledonians, carefully watched, an integral part of their lives.

Not all megaliths are astronomical. More than 100 circles and standing stones in Britain have proven sun and/or moon connections, but for others there is no astronomy to be found, or

The moon stump (marked by stake and man) lines up with a distant peak when viewed from the center of Temple Wood circle.

if it ever existed the evidence has gone. Not all archaeologists accept the findings. Their doubts range from helpful questions raised at conferences to outright rejection. One British authority cannot accept astronomy in the ancient structures because it doesn't agree with what he was taught. Another cannot believe such capability could come from the rough, mud-floored huts of ancient Britain.

I do not argue with these viewpoints. They are rightful personal opinions. Each scholar must decide on the basis of the evidence within the framework of his or her own perceptions. My perception comes from the objective side, not the subjective. At age 16 in the old British educational system I was forced to choose between arts and science, as the split was then called, and with ambivalence I chose the latter for a career. From that point onward humanities and social science were cut out from my education, unless I was going to echo the words of Arthur Quiller Couch and say, "Degree from London University, but largely self-educated." In any debate over the new discoveries I was restricted to explaining the astronomical/mathematical side and could only stand by and take note of the slow development of archaeological/anthropological opinion.

For those few who rejected the theory outright there was of course nothing more to say. One critic flatly said Stonehenge was built by howling barbarians who were incapable of using it as the theory proposed. Two other critics concluded that the builders were herdsmen-chieftains and the 56 Aubrey holes were dug by, or were symbolic of, 56 separate families or tribes.

The question of Neolithic accuracy came up. The accuracy of pointing was usually no better than 1 degree. This was the average at Stonehenge, Callanish, Kilmartin and elsewhere. There were exceptions like the heelstone, which was precise for the revised date as viewed through the sarsen archway, but there was no way of telling in an analysis if the exceptions were not due to chance. The odd bull's eye here and there may not really be significant if the other shots are scattered around it. But one or two experts wanted greater accuracy. If the human eye can see detail to a fraction of a degree, they said, then the Stonehengers should have achieved this accuracy in the stones. Because they did not, the alignments are invalid, or so the argument went.

Personally I couldn't agree with this. We cannot impose our

Long Meg marked the midwinter sunset as seen from the circle, Cambria, England.

standards of accuracy on a Neolithic culture. If the builders accepted errors of a degree, then so be it. Archaeologist-prehistorian Colin Renfrew came up to me at a London conference when the errors were being discussed and said he didn't think the lines were the original observational sight lines. Those archways were "reminders." Stonehenge was really a data bank, a repository of knowledge. Astronomer Fred Hoyle believed the offsets were built deliberately to help the priest-astronomers in their work. He said if the sun or moon went a little bit past the stone in the swing it helped to pin down the precise day of maximum.

Civil engineer Alexander Thom claimed super-precision with accuracy exceeding the resolution of the human eye. He said the Neolithic observers could watch the instant of disappearance of the last flash as the sun went down behind suitable notches on the distant skyline. This might be so, but I personally have doubts. When the alignment is a row of stones, the builders clearly had an alignment in mind, but when the farsight is a natural feature of the landscape, there is a credibility gap. At Kilmartin the "notch" proposed by Thom was so insignificant that I was forced to abandon the idea and concentrate on what was actually there, the anthropic parts. Again at Stonehenge Thom surveyed a small mound on the Avenue skyline, claiming superprecision for the builders, but subsequent excavation showed that the mound was grassed-over modern rubble.

Anthropologist Yvonne Schwartz objected to the term Neolithic science, preferring to call it "magical understanding." She has a point. When I myself, think of science in prehistory I do not link it with quarks, gene splicing or 4-dimensional cosmology. The word "science" originally meant an understanding gained by observation rather than intuition. There may have been a bit of both at Stonehenge. If the word "science" conjures up the wrong image we should not use it. But I don't think "magical understanding" is quite the right phrase either. My own brand of astronomy has tended to be the visual, photographic brand, rather than, say, computer models of exploding galaxies, and when I talk about astronomy in the megaliths I mean the observational. I have tried to emphasize the environmental awareness, the mind-probing observer aspects. When the Stone Age people saw the sun and moon fall in the slots which they with great labor had erected, and the intervening years agreed with the numbers they had set out in the circles, I can't perceive it as entirely magical. What they were doing was more solid than that.

Other anthropologists, Leon Stover and Bruce Kraig, have objected to the words "Neolithic observatory," and "ancient astronomers." In their view an observatory is like the one on Mt. Palomar, and astronomers are white-coated workers who look through the telescope. This is an unfortunate perception. Nowadays astronomers do not look through a telescope, and they prefer casual clothing. The telescopes provide photographic and photoelectric data which are disseminated, shared and discussed. It is not much different from a person standing inside the stone telescope watching the sunrise and coming out to tell the others how it went.

C. P. Snow shocked academia when he dared to say there were two cultures, humanities and science, and neither one spoke to the other. I dedicated my book *Stonehenge Decoded* to his two cultures because it seemed to me the subject matter straddled the gap. Some people say the division is fundamental, almost like the coldly logical left brain and the intuitively irrational right. Snow himself acted in both spheres, the literary and the scientific, but since his time the gap has widened. Scientists have extended the frontiers of the material world and humanists have delved deep into the mind. For some it may be disquieting to see astro-archaeology moving across the gap, encroaching on the legendary past. Yet our modern division did not exist in the nonliterate Neolithic period, or even in the literate Renaissance. Understanding the cosmos was one broad human goal involving the whole brain, the intellect of heart and mind combined.

The largest stone circle in Britain is at Avebury on Salisbury Plain, 20 miles north of Stonehenge. It covers more than 28 acres, roughly circular with a diameter of 1,089 feet—big enough to hold 7 Stonehenges. There are the remains of 2 stone rings inside the larger one, and the whole is within a high chalk bank and ditch which was originally 30 feet deep. The archaeological work is still going on at the site, and the astronomy, if there is any, has yet to be discovered. Avebury was occupied when Stonehenge was in use, and it might have been more important. Only vague clues have survived the near-total destruction of the monument by 17th-century demon-exorcists.

Originally about 100 large stones were set around the inside edge of the ditch. This method of placement accounts for the deviation from true circular shape. The ditch was dug in sections, each section probably being the responsibility of a single family or tribe. The ditch and chalk wall made from the digging were defensive, and people herded cattle into the enclosure across the 4 entrance causeways. With picks made from deer antlers and baskets woven from twigs it must have taken 1,000 person-years to dig the ditch, and more time to haul and erect the stones. The largest of the stones were left intact by the exorcists, and wisely so. When archaeologists excavated a toppled stone in the south quadrant they found the remains of a medieval man accidentally flattened. The exorcists' method was to dig a pit, light a fire to crack the stone, and then smash it or bury it. Apparently it fell too soon for him to get out. The bigger stones weigh upward of 50 tons. Two together would equal the space shuttle Columbia in weight.

Work began at Avebury around 2500 B.C. and the place was in use for several centuries. It was a time when burial customs were changing. Before Avebury, people were interred in the so-called Long Barrow at West Kennet about a mile away. This was a chambered tomb similar to New Grange and close to it in date, 3200 B.C. Like New Grange it was sun-connected: the entrance faced due east, the direction of the rising sun on the first day of spring or fall. Like New Grange it was open to the rays. There was no light box, but the door was open. Evidence of scattered bones shows that the living visited the dead and used the skulls in ceremonies.

Avebury's circles and the disused tomb were later connected with a stone-flanked processional avenue. The avenue was built in sections. A special grave was found at one of the junctions.

A diamond (female?) shape stands by the ditch at Avebury.

The grave goods were foreign to Avebury, and archaeologists believe that a passing stranger was sacrificed to sanctify the work. The stones were alternately oblong and diamond in shape with the one standing across from the other. The oblong is taken as a male symbol, the diamond female. Together the symbols stood for fertility, and combined they stood for intercourse.

By the time of Avebury, the dead were cremated and the charred bones buried in urns or in simple pits in the rings. There were dances, ceremonies and propitiations. The north and south stone rings were ceremonial, in the manner of the recumbent circles of Ireland and Scotland. The enclosure was probably defensive in two ways, (1) a difficult obstacle for human and/or animal enemies to breach, and (2) a sanctified area, closed off and protected from spirit-enemies.

Archaeologists paint a bleak picture of the lives of the Avebury dwellers, and an even bleaker one of what British writer Jean McMann calls the "Neolithic mentality." But were those stone haulers only interested in digging into the earth for whitened bones, did their lives begin and end with cattle herding, was there a ghost to be feared in every stone, and was the chalk embankment the outer edge of their narrow world? To be sure, the material facts show skeletons with arthritis, postholes for meager mud-floored huts, and remnants of hunting, berry picking and a low level of agriculture. But what of the mind, the part not found by the spade. Was the culture firmly set in earth-based thinking, or had those people reached beyond the circle to the cosmos? Was it purely mindstep 0, or was it the beginning of mindstep 1?

No one has yet published an analysis, but there are some clues. The numbers are suggestive. There were 30 stones in the south ring, the integer of best fit to the moon-month, which averages 29.53 days. There were 12 stones inside the north ring, the best fit to the number of "moons" in a year. Excavators are unsure of the number of stones originally in the north ring itself; the number is either 30 or 29. They are uncertain because part of the ring is under the road and houses of the village. Astronomically I hope it turns out to be 29; then the arrangement is the same as the Y and Z holes at Stonehenge. By counting in the north and south ring alternately, the builders would come out with the next best approximation to the moon period, 29½. In our number system they would have progressed from the integer to the first decimal.

There was a "cove" in the center of the north ring—three stones faced toward the northeast like a giant's armchair. Archaeologists believe it was a dummy, or symbolic, tomb. Aubrey Burl figured the alignment to be toward the most northerly moonrise. The armchair giant watched the moon rising there once every 18 or 19 years. Burl says they connected the moon with death, or maybe rebirth, or dead souls. The whiteness of the disc was equated with the bleached ancestral skulls found nearby, and with the white quartz found near the recumbent moonstones of Scotland and at the tomb entrance at New Grange in Ireland.

There used to be a tall menhir near the center of the south ring. It has been replaced with a concrete obelisk. A line from the obelisk to another stone could mark the setting moon at the extreme southerly turning point, −29, but this has yet to be checked out. There is an uncertainty in the height of the original bank, and an astronomical survey is needed. There is the stump of a menhir outside the ring. It does not mark the sun or moon when viewed from the obelisk and seems to be more connected with the line joining the centers of the north and south rings, a line which is sometimes called the "axis."

If any part of Avebury is astronomical this axis should be. It was a carefully laid-out line, because there were plans for a third ring immediately to the north. The bearing angle is 340 and it points to declination 35 to within a degree or so, depending on the bank. The star alpha Cygni is on the line, but I doubt if that was the target because the star is only the 18th-brightest star in the sky, and beyond that are the 5 bright planets and the sun and moon. If the line was reversed it might hit the southerly extreme of Venus, depending on the exact altitude of the skyline, and if so this would lend support to the suspicion of the Venus influence in the circles of Ireland and Scotland. It may again be coincidental, but the Avebury axis has a connection to the Kilmartin stones. At that site, the line of the pairs, the line to the northwest menhir, the faces of the oblong slabs (even the buried stump), all had a mysterious parallelism, pointing to an unexplained declination of 34 degrees. So to within a degree, Avebury's axis and the Kilmartin slabs at that remote time in prehistory pointed to the same part of the sky, a position where there is now no identifiable target. Perhaps it was the edge of the band of the zodiac, the line beyond which no planet could go, or perhaps it was some long-gone explosive object like a supernova.

Someday, I believe, a way will be found to unravel the Avebury mystery. It may be a site where the emphasis was on the earthy, the occult and the somber, and the local culture was at mindstep 0, or there may have been a cosmic component. As at Stonehenge, large sections of Avebury remain undug. The ground is no doubt rich with clues for future research in archaeology, anthropology, and perhaps astronomy.

Avebury, because of its size, has been called the cathedral of the stone circles. Indeed, the name may be appropriate, but to my mind Stonehenge, its southern neighbor, was equally a cathedral and a greater wonder. In architecture and in coded knowledge it was unique. Stonehenge, complete, took about as long to build as the Gothic cathedrals, and was the most intricate and impressive of all the British stone circles. Like those cathedrals, Stonehenge was probably a place of worship, an area for congregating and a focus of learning and knowledge. For we who can only look at it as it is and attempt to imagine it as it was, Stonehenge stands as a monument to the concepts of mindstep 1.

SIX

Pyramids and a Waning Princess

Stonehenge was in its early phase in 2700 B.C., at the time the first pyramid was going up at Giza. Archaeologists used to think that the Nile pyramids came before Stonehenge, but now, with a correction to the radiocarbon dates, they know the building projects were going on simultaneously in Egypt and in England. Not that there was a direct connection. The two cultures were poles apart, different in language, customs and religion, with only the slightest contact, if any. Nor are the civilizations to be compared in wealth or power. But the radiocarbon clock now tells us that the two societies lived at the same time, and the astronomy reveals some of the same underlying motives for the structures that were built. The same stimuli, sun and moon, produced the same reactions. Those archways on Salisbury Plain make a cosmic connection, or so it would seem, and those pointed shapes viewed across the desert haze at Giza also connect. Like Stonehenge, the pyramids are a monument to the mindstep, the first structures to be made from cut and dressed stone, the largest stone constructions in the world—ancient or modern.

How could the pharaoh inspire 100,000 subjects to cut and haul 2 million blocks of limestone over a period of 20 years to make a sloping-sided, north-south-oriented statement in geometry? British physicist Kurt Mendelssohn said the reason was the

NOTE: This chapter includes new results from work in Egypt, 1970–72.

work itself. The work project was devised to keep the population busy during the fallow period between harvest and the next inundation of the river Nile, and busy populations are peaceful populations.

An interesting idea, but I can't agree with the work-project theory as the reason for the pyramids. There was more to it than that. Certainly the organizing of the labor was impressive and would make a fine political-sociological focus, but why the pyramid shape? Why not a cube? Why not a construction of practical use and visible pay-back like an amphitheater or a road system, bridges or palaces? No, there was a compulsion to build that pointed shape visible for miles around, to align its base square to the cardinal points accurately to minutes of arc, and to cover the 4 flat faces with smooth white limestone.

The old idea of whiplashed slaves heaving in agony is probably wrong too. Mendelssohn was right on the socio-economic aspects—feeding and housing of the work force, economic pump-priming, political power—but a task occupying a whole nation and lasting from one generation to the next could not be done by a hostile population, seething in revolt. Nor were the pyramids built by the Israelites in their captivity. That event came a thousand years later.

When the Greeks conquered Egypt in 332 B.C. they were told by the priests that the pyramids housed the bodies of the pharaohs. The theory of royal tombs was readily accepted. It was logical for a great ruler to prepare in a great way for death, and the solid mass of blocks guaranteed protection of his remains. But Mendelssohn in his research has challenged the tomb theory and set in motion a rethinking. True, the pyramids contain passages and chambers, and a few have stone coffins inside. But the bodies are missing. Tomb robbers? No. When the plaster-sealed sarcophagus of Sekhemkhet (approximately 2700 B.C.) was opened, the coffin was empty. Although the pyramids are surrounded by mortuary temples and other signs of the afterlife, they probably were empty when closed off, and the pharaoh's body buried elsewhere. If anything was buried in the pyramid it was his "ka," his soul, the eternal, invisible body.

Today the pyramids have become a numerical, geometrical challenge. I will not go into the claims of pyramid power, or into the meanings attached to the lengths of the sides or the height, nor to the motive of the pyramidologist who was allegedly caught by Egyptologist Flinders Petrie bashing a corner of a stone because it did not fit his theory. These are modern problems. But certainly some aspects of numbers, angles and 3 dimensional planning must have gone into the building of those structures.

It is commonly said that the Great Pyramid of pharaoh Khufu (known as Cheops to the Greeks) reveals a knowledge of the number pi, π. The slope angle is close to 51° 51′, with tangent equal to $4/\pi$. Actually the height before the casing stones were ripped off to make the buildings of medieval Cairo was 481.4 feet, and the length of the bases averaged 755.79 feet, making a slope angle of 51° 52′. This is a good fit to the angle whose tangent is $4/\pi$, and the slope could embody that knowledge. However, the slopes of the 2nd and 3rd pyramids are not as accurate, being 53° 4′ and 50° 43′. The value of π is a very definite, mathematical number, and it is unlikely that the builders went wrong by nearly 10% in the 2nd and 3rd pyramids, as the slope angle would imply. It is more likely, to me, that these 3 pyramids taken together were not built to enshrine this rather tricky number, π, nor do I think the angles resulted from rolling a wheel as Mendelssohn suggests—the angles would have come out much more in agreement than they are.

I wonder whether the angles are related to the circumference of the vaulted heavens. Seven was a mystical number to the ancient world of mindstep 1, representing the 5 planets plus the sun and moon. The circle of the sky was the cosmos itself, the heavens above and the underworld below. It was the protective loop that wrapped around the name of each pharaoh in the cartouche of his royal name. Each pharaoh might have wanted "one-seventh of the heavens" built into that pyramid monument.

The slope angle of the Giza pyramids is very closely one-seventh of the sky circle. This fact has been overlooked until now because of the tricks of numbers. The 360 degrees of our circle (inherited from the Babylonians) can be divided evenly by all the digits except 7. For example, dividing 360 by 2, 3, 4, 5, 6, 8 and 9 gives whole numbers: 180, 120, 90, 72, 60, 45 and 40. But dividing by 7 does not work: the answer in the 360-degree system is 51.428, or 51° 25′ 43″. It would be no matter to the ancient Egyptians that it came out to such an awkward value in degrees, minutes and seconds, because 7 was the number they were

Granite block from the base of the Great Pyramid showing the slope angle.

choosing even though it doesn't show in the degree system. For the mason, 7 capstones pushed together would fit neatly like the spokes of a wheel. As it turned out, the first pyramid angle is 360/6.9, the second is 360/6.8, and the third is 360/7.1. I prefer this interpretation because it would be something the general population could look at and understand in its mystical meaning.

It is commonly said that the descending corridor in the north face of the Great Pyramid pointed at the time of construction to the pole star. This is interesting astronomically, but unfortunately it is a fallacy. The slope of the line misses the celestial pole by more than 3 degrees. True, there was a faint star in Draco circling the pole in 2700 B.C. and lining up with the corridor in the low point of its circuit. But there are many faint stars in the sky and I doubt that the correlation was significant. Nor could it be called a pole star, moving as it did in a 6-degree circle.

I wonder if it is not the corridor but the face angle that is astronomical. The south face of Khufu's pyramid is built to be square-on to the rays of the midwinter sun at noon. Around 2700–2600 B.C., at latitude 30, the sun was 36 degrees above the horizon

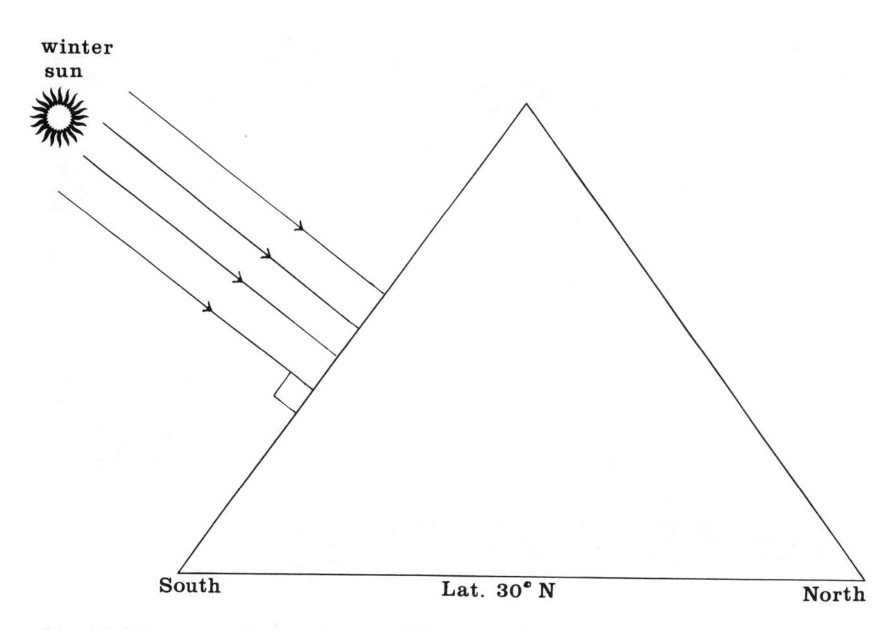

At midwinter noon the sun's rays fell perpendicularly on the pyramid's south face.

in the winter (program NOONALT), within a degree or so of the perpendicular to the face. This means that at the winter solstice the sun shone directly at the sloping surface of polished limestone, giving light and life-giving warmth to the stones. With the 2nd pyramid the fit is even better; the error is less than a degree. If the intention of the builders was to connect with the sun at its lowest traverse, this would be the first known alignment built toward an object halfway up the sky. If the builders chose the sun as a target in this way, then the pyramid shape was dictated by the tilt angle of the ecliptic and latitude, the earth-sky connection. Truly a cosmic design. Were the Egyptians building a mirror so that the weakened sun could see its image and be reborn? Modern earth-artists have signaled into space by reflecting sunlight off large surfaces. Whether by design or accident, the Giza pyramids did the same 4,600 years ago. Each midwinter's day the sunlight was returned upward into space along the very path it had come down.

How did the builders set the base so accurately north-southeast-west, and why did they do this? The error of alignment amounts to no more than 5 minutes of arc, the thickness of a piece of string viewed at a distance of 6 feet. If they took sightings of stars rising in the east and setting in the west, then the mid-point would come out to be due north. Or they could have marked sunrise and sunset on the longest (and/or shortest) day of the year, and the bisector would again give the north-south line. However it was done, astronomical measurements were needed, and they pushed those measurements to the very limits of the accuracy and precision of their day.

Our modern streets run east-west for the convenience of surveyors and map makers. In ancient Egypt the motive was less trivial. There was a desire, an urge, to link earthly structures to the sky with long imaginary lines. East and west, those places on the horizon, were the abode of the blessed, the transfer points between earth and heaven. The sphinx in front of the 2nd pyramid faces due east, and like the pyramid looks toward the spirit world on the far horizon.

With each base placed along lines of latitude and longitude, the pyramid forever afterwards would be a giant gnomon, a functioning, seasonal sundial with a shadow about 1,000 feet long, sweeping over the desert. Unfortunately, if ever there were markers in the desert to count off the Egyptian hours, these markers have disappeared. Or maybe they are beneath the sand, awaiting the archaeologist's spade.

Stonehenge and the Giza pyramids were products of mind-step 1, and there are similarities. Although smaller in scale, Stonehenge was at the limit of the building capabilities of the ancient Britons. It was aligned to the sun, was their first structure of cut and dressed stone, was associated with burials yet had no tomb, and was a focal point for the culture. There was no writing on the stones of Stonehenge, and there is no writing carved on the blocks of the Giza pyramids. (Some blocks have painted signs indicating the name of the construction gang, but the hieroglyphic inscription over the entrance of one of the pyramids was carved, alas, by an Egyptologist of the last century who should have known better.) Stonehenge was a religious symbol like the pyramids, a pseudo-calendar, a place for occult astronomy, and a locale of priests and leaders. We cannot fully know all the details of Stonehenge's purpose without a backward trip in a time machine. There was some unifying idea behind it. In the same way the clues to the motives of the pyramid builders are there if only we could piece them together.

After a life's study of all things Egyptian, Dr. I. E. S. Edwards, Keeper of Antiquities at the British Museum, came to interesting conclusions. He said the pyramid's shape represents the rays of the sun slanting downward as they do when shining through gaps in the clouds. The sloping side was a frozen sunbeam "as a means whereby the dead king could ascend to heaven." Edwards translates the Egyptian word for pyramid, *mer,* as the "place of going up." He quotes magic spells from the later pyramidic scrolls which tell of the pharaoh going up to the circumpolar stars: "Spell 267: A staircase to heaven is laid for him." The pyramid shape, says Edwards, is "the very symbol of the Sun-God." It is modeled on the benben, the mythical conical stone which first appeared above the waters of chaos at the creation of the world. At the instant of the first dawn, Atum, god of creation, settled on the top of the benben stone in the form of a phoenix. This conjures up the image of the sun's disc standing on the tip of the heelstone in far-away Britain!

There is another clue. It comes from the Mesopotamian valley. Herodotus visited the ziggurat of Babylon and wrote:

> On the topmost tower there is a spacious temple, and inside the temple stands a great bed covered with fine bedclothes, with a golden table by its side. There is no statue of any kind set in the place, nor is the chamber occupied of nights by any one but a single native woman . . . chosen for himself by the deity out of all the women of the land. They also declare—but I for my part do not credit it—that the god comes down in person into this chamber, and sleeps upon the couch.*

This is thought to be a reference to the Biblical Tower of Babel. Edwards says that the pyramids, like the ziggurats, were functional, at least in the imagination. They allowed the pharaoh's soul to pass, like a god, from the earth below to the sky above.

Perhaps the pyramids were built as cosmic openings. Living in mindstep 1, those people below were wrapped in a canopy of

* George Rawlinson, ed. and trans., *History of Herodotus,* vol. 1 (Everyman's Library, 1912), pp. 181–82.

celestial gods, and it would be logical to reach up into the sky with the highest structure that could then be built. The builders placed a small black pyramidal stone at the top. It was a magic tip. An ancient Egyptian could easily imagine the sun-god coming down momentarily to rest on the pyramid, like the phoenix settling on the benben stone. Pharaohs and priests could ascend those sun's rays frozen into stone to speak with Ra on high. The humble worker who hauled the stone blocks might believe that just touching those magical sides would siphon off a portion of the power of the cosmos for him. Religion and magic connected from the sky by white stone to the earth!

Astronomers have found other clues in ancient Egypt. The twin statues erected for Amenhotep III (around 1380 B.C.) and wrongly called by the Greeks the Colossi of Memnon were sunrise connected. Until an earthquake dislodged one of the stones, the statues gave out a faint, eerie cry at dawn, according to ancient writers. The statues and the temple that stood behind them pointed to the sun at declination −24 at the date of construction. Clearly, as with Stonehenge and the New Grange tomb, midwinter sunrise was a significant calendric date to the

The Colossi of Memnon looked toward the midwinter sunrise.

Egyptians. This was because their civil calendar was not fixed with the seasons. It contained a straight 365 days with no leap years, slipping 1 day every 4 years, 1 whole season every 300 years or so. Nevertheless, the time of midwinter was marked by sunrise statues and was celebrated with a festival. The ancient Egyptians had many feasts and festivals controlled by the annual journey of the sun along the zodiac and by the phases of the moon.

In connection with the calendar I was asked to look at Abu Simbel, the largest temple ever to be cut from solid rock, on a cliff on the west bank of the Nile south of the tropic of Cancer. It was built in the reign of Rameses II in the 12th century B.C. Egyptologists wondered if the axis, penetrating 200 feet into the darkness of the sandstone, was designed with a cosmic purpose. On October 17 (or 18 depending on how the leap years fall) by our modern calendar, a shaft of sunlight goes to the far end of

The rock-hewn temple of Abu Simbel with the small solstice chapel to the right.

the temple at dawn, illuminating three statues—Ra-Hor-Akhety (sun god), Rameses II, and Amon-Ra (god of Upper Egypt).

October 17 is not an astronomically significant date, and I asked the experts why they wanted an astro-archaeological test. They thought the alignment might be for a particular date in the slipping ancient Egyptian calendar, the day of a Jubilee celebration of the pharaoh. By custom a pharaoh celebrated the 30th year of his reign with a Jubilee, a day when he was supposed to be reborn. The day was usually day 1 of month 1 of season 2 of the civil calendar. Since season 2 was called Peret, the date could also be written "1 Peret 1." Because Rameses reigned for a record-breaking 67 years, he had two main Jubilees, and he had other Jubilees not in the 30-year system. These ritual renewals became more frequent as Rameses became older.

Astronomically the Jubilee question was a challenge. Because the Egyptian calendar lacked the leap year, the alignment would function as a sunrise flash on 1 Peret 1 for a period of no more than 20 years. The alignment would require accurate site measurements. Ironically, in this century the world's largest temple cut from one piece of stone was monolithic no more. In the 1960s Abu Simbel had been cut into separate blocks and reassembled on the top of the cliff to escape the rising waters of the High Dam. Fortunately the axis of the temple had been preserved and the azimuth was known. Only the contour of the skyline was different. At the original site the sun came up in a notch on the horizon, and in the clear desert dawn the shadow of the distant hill fell on the temple facade. A pink tongue of sunlight slipped down the front of the temple and entered the doorway. We made an allowance for this change, but, short of draining Lake Nasser, measurements at the original site were not possible. The data, however, were accurate enough to answer the Jubilee question.

Vital statistics of the temple are: latitude, +22.1; height above sea level, 123 meters; original skyline, 0.5 degrees. The latest chronology gives Rameses II's first year on the throne as 1304 B.C., and so the 60th Jubilee fell on 1 Peret 1, 1245 B.C., or day 1, month 1, season 2 by the Egyptian civil calendar.* The relevant programs give the following values: EGYPTDATE, −1243.1759; SUN-

*_Cambridge Ancient History,_ vol. 1, part 1 (Cambridge, England: Cambridge University Press, 1970), p. 189.

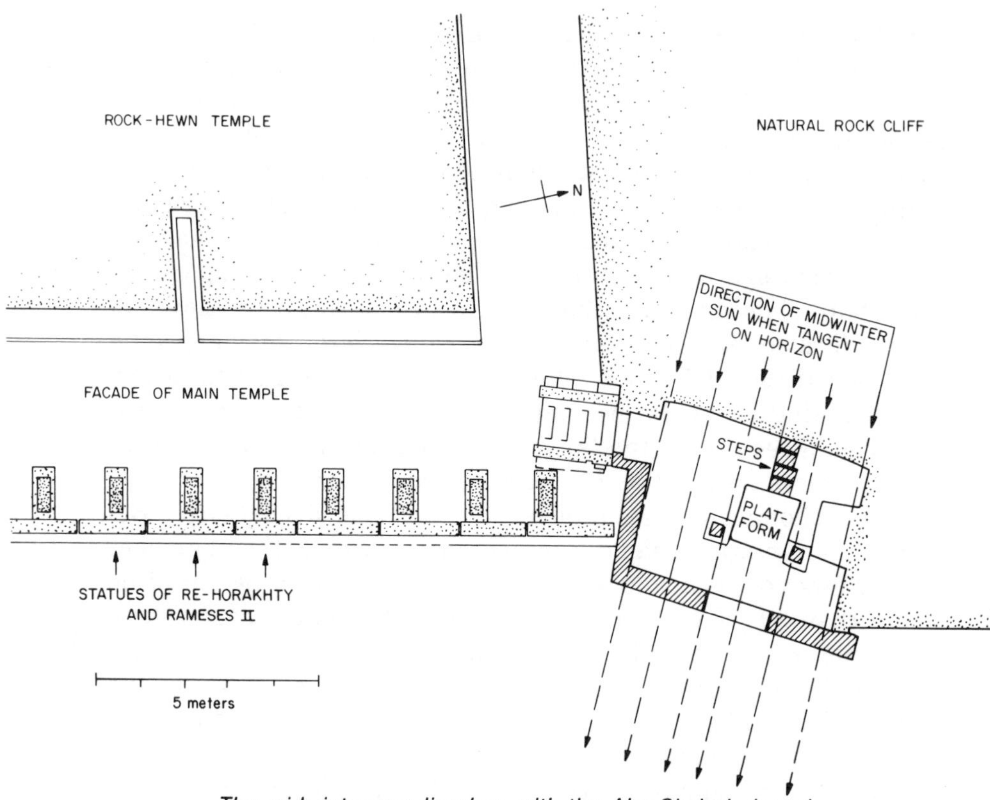

The midwinter sun lined up with the Abu Simbel chapel.

MAX, 23.861; LONGITUDE, 204.05; SUNDEC, −9.488. If we test for the sun standing tangent on the horizon on the Jubilee day, then $d_{obj}=-9.488$, $q=-0.267$, $p=0.002$, and STONEHENGE shows a vertical error of +0.37 and an error in declination of +0.16 degrees. And so with all this calculation the sun comes out to be within a fraction of a degree of the line to the doorway on that date, and the suggestion is validated.

But I wondered at such a gigantic temple being designed and built for a sunrise on one day in that drifting calendar. I wondered if there wasn't a second purpose, an alignment with the seasonal calendar that would be functional year after year, something connected with October 17/18 in the steady, nondrifting calendar. My wife suggested it was related to the inundation, the seasonal flooding of the Nile. Before the time of the last inundation, when the Aswan High Dam was built and the site of the Abu Simbel temple disappeared under Lake Nasser, the Nile waters reached their peak in late September or early October. It is just possible that with the seasonal stability of ancient Egypt the peak of the flood was marked by that sunrise. The waters would recede when a shaft of light from Ra entered the doorway and illuminated the statues deep inside the rock.

Nothing in the Egyptian hieroglyphic writing has been found to explain this October sunrise phenomenon. Astronomically it is perplexing because it is unique, a special date, not in itself astronomical. Solstices are less perplexing and easier to understand in terms of a mindstep 1 culture.

Outside, at the front of the temple, I found a solstice marker in the small chapel to the right of the entrance. This chapel pointed as precisely as could now be determined to the midwinter sunrise at the time of Rameses II, and it parallels the axis of the Colossi of Memnon, allowing for the difference in latitude between the sites.

This little chapel was decorated with a carved tableau showing the barque of the sun-god, and originally contained sculptured statues of baboons—animals traditionally associated with the sun because they howled just before dawn. The main temple was also decorated across the top with a frieze of baboons, and over the entrance doorway, between the oversized statues of Rameses, there was an effigy of Ra-Hor-Akhety. In the entranceway itself, the falcon of Ra-Hor-Akhety was deeply carved in the stone. There could be no doubt of the sun connection of both the main Abu Simbel temple and the side chapel.

The Amon-Ra temple at Karnak also points to the winter sun. Every pharaoh worthy of the two crowns of Upper and Lower Egypt lavished gold and silver on the extension of this structure, building and rebuilding it, yet each monarch maintained the cosmic line. The temple is on the east bank of the Nile in what was ancient Thebes on the outskirts of modern Luxor.

Ra is written as a simple disc in the hieroglyphs, and Amon as a feather, plus a gaming board and a wavy line. Scholars cannot be certain of the ancient pronunciation. The disc could be Ra or Re. The feather is a sound somewhere between our "a" and "ee," the gaming board is "m," and the wavy line, originally connected with the idea of water, is "n." Hieroglyphic writers generally dropped the vowels, so the "o" in Amon is a scholastic "best guess." It could be Amun, Amin, or variations. Some Egyptologists identify Amon as the so-called unseen god. Certainly Amon

(Above) The sunrise god, Ra-Hor-Akhety, was carved in the rock face over the Abu Simbel door. (Opposite) The Great Temple of Amon-Ra pointed toward the midwinter sunrise.

was the most powerful god of Upper Egypt for centuries, and when joined to Ra, the sun-god of the Delta of Lower Egypt, the combination was formidable.

This temple, the largest in the ancient world, was 1,860 feet in length from the southeast gateway to the quay on the banks of

the Nile reached by the avenue of sphinxes. The 12 large columns of the Hypostyle Hall were 80 feet high and 11½ feet in diameter at the top. On each pillar there were inscriptions and scenes showing the pharaoh and gods. At the time of Akhnaton and Nefertiti (around 1370 B.C.), the temple was dedicated to Aton, the very disc of the sun, the hot, pre-eminent life-giving force.

I made measurements at Luxor in 1971. So far as I could tell from site plans and horizon surveys, the axis of the temple pointed to the disc standing on the horizon at midwinter's dawn at the time of the major reconstruction around 1480 B.C. The alignment was accurate to better than one-tenth of a degree. A team from California's Griffiths Observatory took a photograph in 1978 which, allowing for precession, confirmed my calculations.

On top of the northern wall stands a small roof temple, the High Room of the Sun, dedicated to Ra-Hor-Akhety, the sun-god of the two horizons. (Akhet is a feminine word meaning brightness, or horizon, and the "y" at the end signifies a double plural—thus "of the two horizons.") The two horizons could be the two extreme rising positions, midsummer and midwinter, +24 and −24. The temple was open to the sky—presumably for star observations—and a slot in the southeast wall was conveniently placed for sunrise watching on that critical day.

The roof temple stands directly over a wall inscribed with details of the foundation ceremony. Pictures show the pharaoh staking out the survey cord, hoeing the foundation line and laying the bricks. On another wall the roof temple is referred to as "aha," meaning "place of combat." Maybe this refers to the sun overcoming the forces dragging it down at its lowest point on the ecliptic when, unlike mortal Gilgamesh, it is born anew. More poetically, the aha is described in the same inscription as "the lonesome place of the majestic soul, the high room of the intelligence which moves across the sky."

Thebes had a trinity of gods—Amon-Ra; Mut, the wife of Amon-Ra; and Khonsu, the god of the moon. Khonsu also takes the form of a child, a wanderer and a smiter of wandering demons. All three have large temple complexes, and these are built to three different rectangular grids. The Mut complex is south of Amon-Ra, reached by an avenue of sphinxes, and is now very much in ruins. The Mut temple was set in a crescent-shaped sacred lake. There was a parallel temple to the west and two trans-

The stairway leading to the High Room of the Sun, Amon-Ra temple, Karnak.

The ruined temple of Mut stands in the crescent of the sacred lake.

verse temples to the north of the Mut temple, all built on a grid rotated by an angle of 8 degrees to the grid of Amon-Ra.

Khonsu's main temple is inside the mud-brick boundary walls of Amon-Ra, built on a grid rotated by 1.9 degrees. Work on that temple was begun by Rameses III, the last great pharaoh to bear the Rameses name, at an approximate date of 1180 B.C. There was a small temple on the roof, open to the sky, with a window facing transverse to the main temple axis, across to the hills of Thebes on the west bank of the Nile. An inscription is still visible over the window, although now somewhat damaged. It says: "Live the good god, beloved of Amon [text damaged] who shines on the benbenety like the horizon house of [text damaged]." Astronomically, one of the two "bens" could be the turning point of the moon on the western horizon, and the moon god could be the new crescent. The 1.9-degree rotation put the new crescent exactly in the window, with declination +27.4 at midsummer. (Program STONEHENGE, azimuth 298.8, latitude 25.7, height 77 meters, skyline 2.6.)

The axis of the Khonsu temple, as extended by later pharaohs, lines up with the setting of the star Canopus, known astro-

Pharaoh and moon-god Khonsu ( ) in the Khonsu temple.

Images of the "old moon in the new moon's arms" carved in the Khonsu moon temple, Karnak.

nomically as alpha Carinae, the 2nd brightest star in the sky. The axis points along the Nile valley where the skyline is low, 0.0 degrees, and the target declination for a star is −52.6, close to Canopus at −53.5 (program STONEHENGE with 0 diameter and parallax since the target is a star). Stonehenge was based on a right angle, the station stones marking the 90 degrees between extreme positions of the sun and moon. The pyramids also have square corners, and the edges mark the sun at equinox and noon. Here again with the Khonsu temple we have a structure incorporating a right angle of the cosmos, and in this grid it is to the northerly moon and a bright star. Canopus could be the second "ben." One could, of course, argue that these things occurred by random chance and that the builders were in ignorance of the way their foundation lines matched the sky-gods, but I believe the weight of evidence is otherwise.

For example, when the moon first shows at the beginning of the month, age 1 day, it is a thin crescent. At the low latitude of Thebes it would seem to be lying on its back. The old moon glows faintly in the crescent as earthlight falls on the darkened

Hieroglyphic inscription by the moon slot in the Khonsu roof temple.

moon. When the Atlantic and Indian oceans are cloud-covered, the reflected sunlight is sufficient to reveal the features of the full moon as a dim, pale orange orb—Enkidu held in the woman's cusp-embrace. This is exactly the way the Khonsu-god is shown in the temple. Carved over the doorways and on the walls we see the full disc held inside a thin crescent.

Khonsu as orb and crescent was in the window slot in the month of midsummer when it was new following the solstice. That happened once every 18 or 19 years! The slot was wide enough to show the crescent in the year before and the year after, and during the year the full moon and other phases would take up position, but the extreme (the ben?) was fixed by the 18.614-year cycle! At that time the moon was setting on the crest of the Theban hills in the direction of the famous valley of the kings, the hidden tombs of the dynasties from Tutankhamen to Rameses.

What about the other turning point 9 or 10 years later when the crescent moon at midsummer is farthest south? This was where the grid of Mut came in, set within the crescent lake. Year by year the summer crescent moved along the outline of the hills of Thebes, rested and then returned. So many things in the grid picked out this direction—the horns of the crescent lake, the

Axis of a ruined moon temple in the Mut grid. It points toward the Theban hills in the direction of the Valley of the Queens, and the lowest setting point of the northern crescent moon.

transverse axis of Mut, the long axis of the temples to the north of the complex. One of these temples was dedicated to Khonsu the child, as if the moon god was born again at this low point. Mut was the wife of Ra, and in some accounts the moon was their child. Interestingly enough, this magic line to the new moon strikes the hills on the west bank at the area of the tombs of the queens. The northern extreme was for the kings, the southern for the resting place of their queens.

Now dynastic Egypt was a literate society. The hieroglyphs have been deciphered, thanks to the brilliant work of French linguist Jean Francois Champollion and British physicist Thomas Young, and we have actual written accounts from that distant age. But in reading the texts we enter another world. The calendar is more or less understood. The Egyptian civil calendar contained 3 seasons each with 4 30-day months, and 5 extra days were added at the end to make it 365—no more, no less. It seems that the priests never lost a day in their counting until the empire was taken over by the Greeks and then the Romans. Censorinus, writing in 238 A.D., noted how the 1st day of the 1st month of the 1st season fell on July 20, 139 A.D., by the Old Style calendar of Julius Caesar, and so we can trace back through the millennia to dynastic times (program EGYPTDATE).

But there is a vast area of Egyptian writing that is difficult to understand. Maybe it was deliberately secretive. The area of death, religion and myth is almost unintelligible to us, even though we can translate every word. The sun was a golden bird, or a boat sailing across the sky, or a scarab beetle rolling its egg in a ball of life-giving dung, or it was a disc of gold and/or the body of the pharaoh at one and the same time. Apparently all things could be believed, even if there were conflicts. Generalizing, the process of connecting ideas, was a step away. The power of abstract thought was to come later. One pharaoh, Akhnaton, was different. He introduced worship of the sun's disc, a precursor of monotheism. With one of the largest brains of prehistory (judging by the cranial cavity in the skeletal skull), Akhnaton was grasping at a new approach, but the priests could not, or would not, follow. The crime of Akhnaton in that ancient Egyptian environment was to rationalize. He perished, and the meme went with him.

Pharaonic Egypt spoke and wrote in spells and stories. They were recited, remembered and written down. For want of a better word we call them myths, but the content was almost certainly believed as truth by those people, inconsistent though the story line might be. And the texts were interlaced with facts, numbers, and pseudo-explanations of natural events. How does the sun move? It is carried in a boat. How does the pharaoh reach the gods? He ascends the side of the pyramid. Why did he drink 7 goblets of wine? Because there are 7 wandering planets. Make note of the meaning of the number! I remember Giorgio de Santillana, humanities professor at the Massachusetts Institute of Technology, stressing that fact in a lecture. He theorized that legends were a vehicle for transmitting numbers from one generation to the next.

Moon-god Khonsu has given us one of the strangest stories to come out of Egypt. In the early 1800s, Champollion inspected a small carved stone tablet which had been found by his coworker Rosellini in a chapel near the main Khonsu temple. It carried in 36 lines of compressed hieroglyphs the legend of the waning princess.

In recent years the stela has fallen on hard times. It was locked away in the Louvre Museum, and because some of the hieroglyphs follow the style of writing of later scribes, and because experts had difficulty in making sense of it, the whole story was regarded as an invention, a pious forgery of latter-day priests to inject authenticity into a declining moon religion. But earlier scholars took the stone tablet to be a faithful copy, maybe from a papyrus scroll, of an original story, "The Legend of the Waning Princess," passed down from the earliest days of greatness of the Khonsu temple. The story deserves a second look. It gives an intriguing glimpse into the mind of mindstep 1.*

* This text is the author's translation, with the assistance of Dr. Labib Habachi, and paraphrasing of previous translations, including:

Records of the Past, Egyptian Texts, vol. 4, ed. S. Birch (London: S. Bagster Sons, 1875).

Ancient Records of Egypt, vol. 3, ed. James H. Breasted (Chicago: University of Chicago Press, 1906).

Ancient Near Eastern Texts, trans. J. A. Wilson, and ed. James B. Pritchard (Princeton, N.J.: Princeton University Press, 1955).

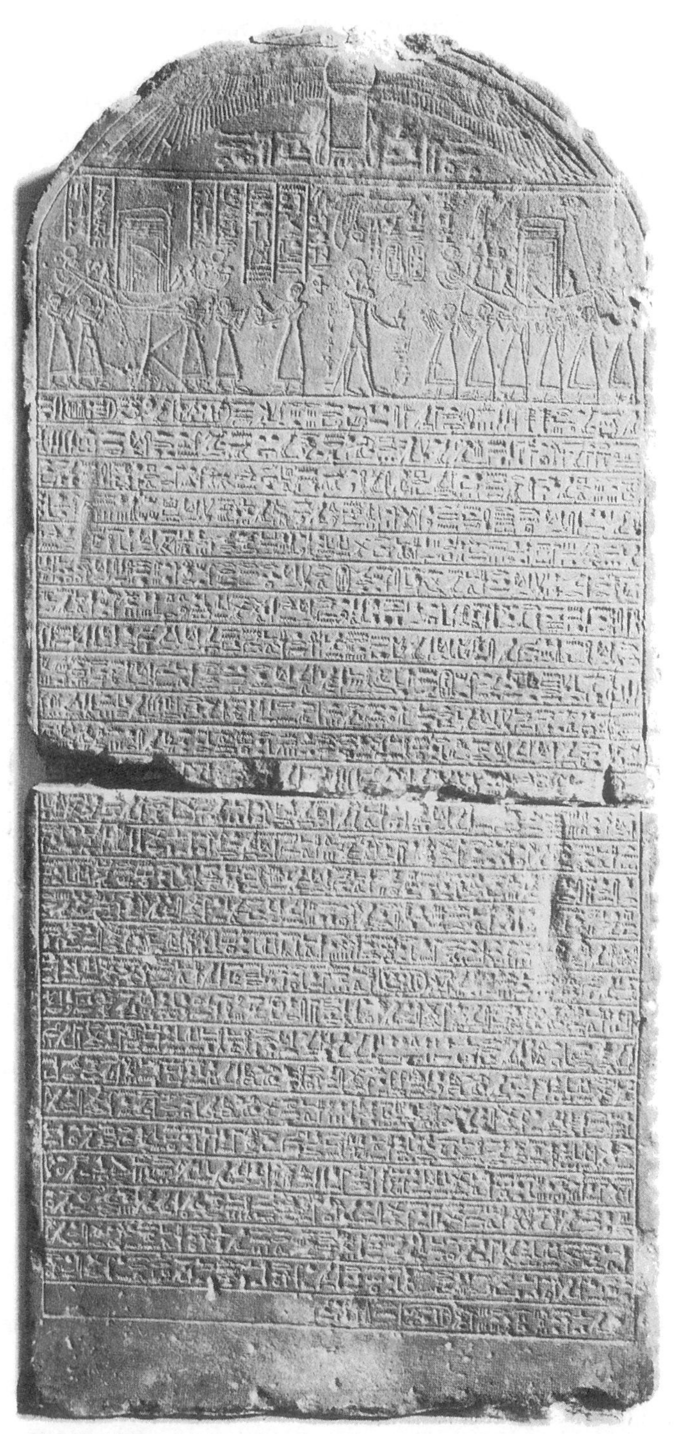

The Bekhten tablet with the story of the waning princess.

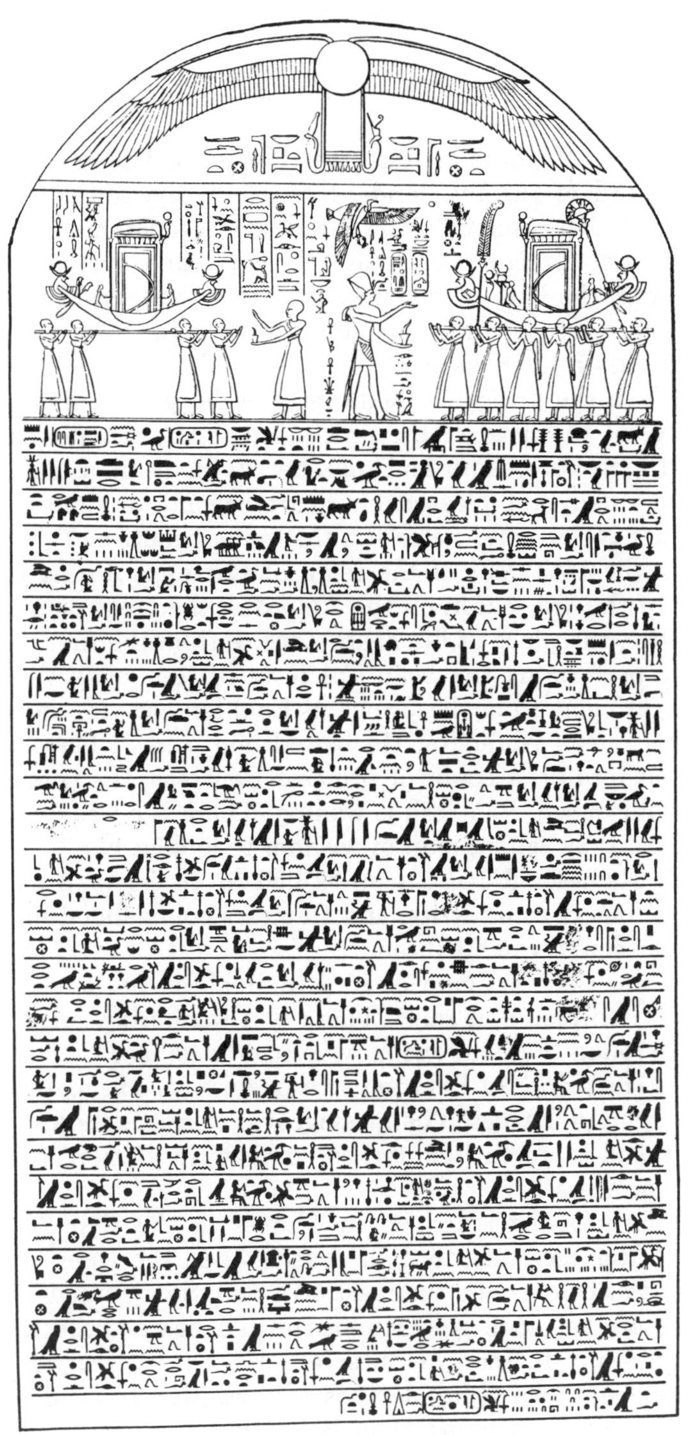

Hieroglyphs on the Bekhten tablet.

★ ☆ ★

"The great pharaoh Rameses II, king of Upper and Lower Egypt, Lord of the two lands, lord of all the gods of Thebes, repeller of the 9 enemy's bows, was in Mesopotamia to receive tribute from foreign princes as was his yearly custom. All the chiefs from the land of marshes and from the land of the rising sun came, each leading his fellow, with gold, lapis lazuli and sweet wood on their backs.

"When the prince of Bekhten came he placed his beautiful daughter at the head of the procession. The great pharaoh was pleased with her beauty and took her to his heart as chief royal wife. The daughter pleased his heart beyond all things and he called her Nefer-nefer-nefer-Ra, the many beauties of the sun.

"In the 15th year of the reign of Rameses, during the 2nd month of the 3rd season on the 22nd day, the great pharaoh was in Thebes, the place of his heart's desire from the first of times. He was performing the beautiful feast of the valley when lo a messenger came from the prince of Bekhten carrying many presents for queen Nefer-Ra. He bowed and said: 'I have come to you great lord on account of princess Bentresh, the younger sister of your queen. Bentresh is waning. There is a weakness in all her limbs. Pray send a man of wisdom to look at her.'

"So Rameses found a man skilled in his heart, a man who with his fingers could write, and Rameses commanded him to go to the far country of Bekhten.

"When the wise man arrived in Bekhten he looked at the princess Bentresh and said: 'There is a wandering spirit within her.' He found there an enemy who had to be faced.

"So the prince of Bekhten sent again a second messenger to the pharaoh in Thebes. The message was to ask Rameses to command a god to go to Bekhten to combat the demon inside the waning princess.

"This messenger arrived in Thebes in the year 26 of the reign of Rameses, month 1 of season 3, at the time when his majesty was celebrating the feast of Amon. So up stood Rameses in front of the moon god, Khonsu in Thebes of beautiful rest, and asked him to send his alter ego Khonsu the carrier out of plans, the god who fights demons who wander, to send this god to the daughter of the prince of Bekhten.

"There was much nodding of the god, and Khonsu in Thebes of beautiful rest nodded and made magical protection 4 times for the god Khonsu, smiter of wanderers, to go to Bekhten and save the princess. He was commanded to go in a ship with 5 boats, with a chariot and horses of the east and west of the moon.

"This god arrived in Bekhten after a journey of 1 year and 5 months. The prince of Bekhten with his soldiers and officials came to the god and lay before him on his royal belly, asking him to be merciful. Khonsu, smiter of wanderers, went to the house where the princess Bentresh was. He made magical protection for her and she became beautiful immediately.

"Up got the spirit who was in the princess and said to Khonsu: 'Welcome, O smiter, O great god, Bekhten is to you a home, the people are for you, and I am your slave. I will go back to the place from where I came to set your heart at rest because of your journey here. But, O smiter, first order a festival with me and with the prince.'

"Khonsu the smiter nodded to the great god and commanded the prince of Bekhten to make a larger offering to the demon that was within the princess. And the prince and his army with him were sore afraid.

"Bekhten's prince held a festival with many offerings before the demon and the god of Khonsu, and the demon left the girl and went in peace to his chosen place by command of Khonsu. And the prince and the people of Bekhten were very happy.

"Then the prince of Bekhten, dreaming of ropes, schemed in his heart to make the god tarry in Bekhten and stop him from returning to Egypt. The god stayed 3 years and 9 months.

"One night, while the prince was sleeping on his bed, he saw Khonsu the smiter come to him outside of his temple. The god was in the form of a falcon of gold, and he flew up to the sky and off to Egypt. The prince awoke, paralyzed in horror. He said: 'Let the god go back to Egypt, let his chariot go to Egypt.' And the prince let Khonsu the smiter return with tribute of every good thing, and with soldiers and horses.

"They arrived in peace in Thebes, and Khonsu the smiter went to the temple of the moon god, Khonsu in Thebes of beautiful rest, to deliver all the good things which the prince of Bekhten had given him. But he did not deliver the whole of the tribute into the house of Khonsu. And Khonsu, the god who fights the demons who wander, he arrived successfully at his own house in year 33, month 2, season 2, on the 19th day in the reign of Rameses II, King of Upper and Lower Egypt, who gives life, like Ra the sun, for ever."

★ ☆ ★

An interesting legend, haunting and full of the ancient Egyptian mystery and life-force. Dr. Labib Habachi, former Inspector of Monuments of the Upper Nile, first showed me this story. A birthright Egyptian who spoke not only Arabic, French, German and English fluently, but also Coptic—that rare language descended directly from ancient Egyptian—he gave me a seminar in hieroglyphics and the ancient Egyptian language. With that I was able to follow with him the symbols and the translation. "What is behind this story?" he asked. It looked to him as though it might be more than a story, perhaps cosmic. Like the legends of Gilgamesh and Quetzalcoatl it is a record of travels, a chain of events involving mortals and celestial gods. There are definite dates, exact time intervals, magic numbers and supernatural happenings.

Could the story allude to the shifting of the moon month by month along the low contours of the Theban hills? Khonsu's main temple marked the northern extreme, and the tablet was found by Rosellini in a chapel at that spot. The story spanned a period of 18 years, from the 15th to the 33rd year of Rameses' reign. Was all the journeying related to the moon in its 18.6-year cycle?

I spoke to John A. Wilson, the late director of Chicago House, the research center at Luxor, about it. Several experts had published their translations of the legend of the "possessed," or wan, princess, and Wilson's was the latest and most accurate. He said: "That legend is a good story, nothing more. Those ancient Egyptians were good storytellers." But I wondered. Even Wilson admitted it had a strange aura, and he said the dates of events "seemed to be carefully constructed."

There seemed to me to be an academic glitch with the starting date. Four experts, Samuel Birch, E. Ledrain, E. De Rouge and E. A. Wallis Budge, translated it as the 15th year of the reign, but Wilson and the famous James Henry Breasted gave "23."

Now numbers should be easy to read. The ancient Egyptians

The alleged typographical error.

wrote 10 as a loop, ∩ , and 1 as a stroke, ı , so there could have been a mistake by the translators owing to the numbers looking alike:

23 is written ∩∩III, and 15 is written ∩IIIII

Before sticking my astronomical neck out it was vital to investigate the glitch. Year 15 would make sense for the moon cycle; year 23 would not.

The stone is now in the Louvre in Paris, and so I called a friend there who visited the museum and obtained a photograph. The stone is now cracked across the middle and worn by time, but the hieroglyphs are still readable. At the top, framed under the protective wings of the sun falcon, there are pictures of the pharaoh and of the two moon-gods in sky-skimming boats. Khonsu the smiter is on the left, carried by 8 attendants, and Khonsu in Thebes of beautiful rest is on the right, borne by 12 attendants. The identifying symbol for "smiter" is the aggressive man-shape to the left of the boat's canopy.

I looked at the photograph of the stone through a magnifier and read off the symbols. Line 6 of the crumbling carving showed 2 strokes and not the upturned loop! The tablet said: "Year 15, month 2, season 3, day 22." The date could have astronomical meaning. I set up the programs EGYPTDATE and LONGITUDE, and with the computer went back in time.

If Rameses took the throne in 1304 B.C., then with "year 15" the scribe had started with April 28, 1289 B.C. My calculator buzzed and flashed—on that day the moon was full, the ascending node was at the vernal equinox, the sun had just passed the vernal equinox on the 1st day of spring, and it was the 1st full moon after the vernal equinox (like the ancient Passover moon, or the modern Easter moon!). The program outputs continued—it was the year of the high moon, MOONMAX was 29, and the moon at Thebes was at its northern extreme, the beginning of an 18.6-year cycle. In that year of 1289 B.C. the summer evening crescent would have shown in the window of the Khonsu roof temple. From the astronomer's viewpoint that priest-scribe had picked a very meaningful opening date!

I checked back through the literature and found that the questionable "23" had originated with one Adolf Erman way back in 1883. In an obscure paper he argued how the 4 journeys and the 1 "stay" add up to a total of about 10 years, and so if the legend finished in the 33rd year it must have begun in the 23rd. He blamed the priest who carved the stone for a "typographical error". In his published copy of the text, Erman altered the II to ∩, generating a real typographical error! In retrospect I must say his argument was weak, his alteration of the copy unjustified and unpardonable.

No, those journeys were not earthly journeys, they were celestial journeys. The numbers are not approximate, they are exact. We are not dealing with a dimly remembered past, we are dealing with a precise record.

No earth journey would take a year and five months! By foot, 10 miles per day was a reasonable journey; by camel or river barge it was 50 or more. Pharaoh Tuthmose III marched an army to Syria, put down a revolt, captured 7 kings and returned to Egypt laden with booty all within a year. Nor has the town or country of Bekhten ever been located by archaeologists. It remains a place of myth and legend—the place where the moon-god had a temple, where the queen's family lived, and where her beautiful sister was possessed by a wandering spirit within her limbs. Those journeys were flights of the gold falcon of the mind.

What about the ending of the legend? The date would be

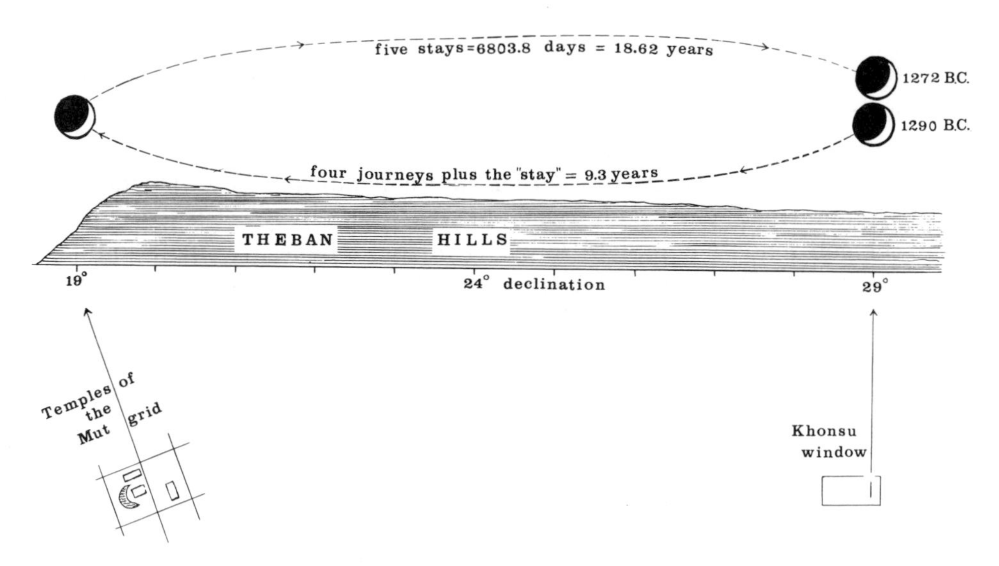

The moon and the princess story.

1272 B.C., December 22. On that date the moon was back in the same part of the sky from which it had started. Modern computers cannot take the moon back with pinpoint accuracy to the time of Rameses II because the constants of the moon's orbit are not known precisely and the spin of the earth has varied enough to generate an uncertainty of 5 degrees or so in the apparent position, but the equations are accurate enough to show that the moon was in the constellation Scorpio on both dates. It is as though the Egyptians were clocking the movement of the moon using a certain star (Antares?) as the marker. We know they had 36 decans along the zodiac. Maybe Antares was one of them.

A simple way to check my finding is to count the days. The moon takes 27.32166 days to go around the zodiac, starting and finishing at a star. This is the "sidereal month." Between the beginning and ending of the legend, 6,447 days have elapsed by the Egyptian calendar, which is 236 sidereal months to within a day. (In practice the moon can be a day late or a day early, depending on how the ellipticity of the orbit and other perturbations affect its motion.) So with the passing of an exact number of sidereal months the moon has gotten back to its starting place among the stars.

For some reason no day number is given for the events in

the 26th year. If it means literally day "nothing," or zero, then it is the last day of the preceding month and the moon again is near Antares—3,963 days have elapsed, or 145 sideral months.

As if to emphasize this fact, the journeys in the story are themselves an exact number of sidereal months. Four journeys are identified: the man of wisdom went, the envoy returned, and the god made a complete round trip from Thebes. A single journey took 512.65 days—1 year of 365 days and 5 lunations. The 4 journeys together took 2050.6 days—75 sidereal months. The 4-fold magic applied by Khonsu also emphasizes the power of the number 4. The travelers on earth passing to and fro between Thebes and Bekhten behave time-wise like the moon and the stars. When the 4 journeys are all completed, they bring the moon back exactly to the same constellation.

What of the "stay," the delay of 3 years and 9 months forced on the moon-god by the prince of Bekhten—is this astronomical? Yes! The interval records the passing of draconic months, the time taken for the moon to go around the zodiac, starting and finishing at one of the nodes (27.2122 days). They are so named in honor of the mythical dragons who lived at the two points (nodes) in the heavens where the moon crossed the ecliptic. A more popular name would be "eclipse month." Three Egyptian years of 365 days and 9 lunar months, the duration of the stay, equal 1,360.7 days. During that interval the moon has crossed the ecliptic exactly 100 times. Khonsu had combatted Humbaba, the dragon at the ascending node, 50 times. With this 50-fold correlation, and we have to note that the number 5 is featured in the story, the Egyptian value for the eclipse month comes out to be accurate to better than an hour. A most remarkable precision, one that could only be achieved by years of watching, noting and remembering.

The length of stay equaling 50 eclipse months might be brushed aside as a coincidence, but there are other factors. Imagine 5 of these stays floating end to end in the zodiac, like the 5 boats that accompanied the moon god. Five stays make up 250 eclipse months, and this, remarkably enough, makes up one complete moon cycle of 18.6 years, during which the nodes have traveled once around the zodiac.

What about the four-journeys-plus-one-stay, the time span in the story that Erman was concerned about? This interval adds up to 9.34 years—one-half of the cycle, the time taken for the extreme moon to move along the hills of Thebes from the Khonsu window to the lines of the Mut complex, from the high moon to the low moon condition. At the start it was setting over the valley of the kings; at the finish it was setting over the valley of the queens. It can hardly be coincidental that this slow movement from high to low is connected with a waning and a sickness, a malady that must be cured. Did they see the shape of the woman in the moon—the harlot of Enkidu—did they see her as the princess of Bekhten?

The Stonehengers with 56 Aubrey holes representing 3 moon cycles were accurate to 3 parts in 1,000. This was arrived at from crude earth-horizon measurements using posts and menhirs. The Egyptians with 5 boats and tarryings were accurate to 1 part in 1,000. Their period was arrived at from zodiac observations and comes out to be 6,803.8 days, a remarkable achievement. In fact we have no way of extrapolating our modern value of the length of the moon cycle back to that era. The Egyptian value may have been the true value of that time. Interestingly enough the Stonehenge and Khonsu values are both slightly greater than the present-day one, so the interval may have become shorter over the 4 millennia.

We can apply an internal test on the numbers in the princess story. If the 5 "stays" did clock off the 18.6-year cycle and the moon was back at the same place among the stars as required by the astronomy, then a whole number of sidereal months must have taken place. Correct. The interval of 6,803.8 days is 249 sidereal months, accurate to within a few hours! Not only has the moon completed 250 draconic months, it has completed an exact number of sidereal months. After a long journey the wandering moon and the node are back again in the same place among the stars.

As an astronomer I am impressed by what the legend conveys. No astronomer, ancient or modern, has ever pointed out the neat arrangement that 250 eclipse months make up 1 moon cycle. An amazing relationship, and entirely an Egyptian discovery. The fit is almost magical—500 weavings of the moon up and down across the ecliptic and then it returns to the starting point. The hieroglyphic sign for magic on line 15 of the tablet is . Translators have taken this to be a rope, a protective knot for hobbling the legs of cattle. Is this a dramatic way of showing the weaving of the moon in the band of the zodiac?

The Numbers in the Legend

Beginning date of the story	1289 B.C., April 28	Year of high moon. Day of full moon. First full moon after vernal equinox.
Middle of the story	1278 B.C., March	Moon near low. Total eclipse of moon Feb. 27, 1278. 145 sidereal months after beginning.
Ending date of the story	1272 B.C., Dec. 22	Moon high. Moon at node. 236 sidereal months after beginning.
Length of 1 journey	512.65 days	
Length of 4 journeys	2,050.6 days	Equals 75 sidereal months.
Length of the "stay" in Bekhten	1,360.7 days	Equals 50 draconic months.
Duration of the story	6,447 days	Equals 236 sidereal months.
Four journeys plus the "stay"	9.3 years	One-half of the moon cycle.
Five "stays"	18.6 years	One complete moon cycle. Equals 250 draconic months exactly, and 249 sidereal months.

What of the nodding? Sometimes the nodding of an effigy was manipulated by the priests of the temple to impress the viewers, but in this story I believe Khonsu in his temple was only imitating what the moon did in the sky. The disc of the moon appears to tilt toward the earth in one section of its orbit, and away from the earth in another. The imaginary face nods up and down and sways from side to side each month. During the 18 years of the legend of the wan princess there were many nods of the head, 250 of them!

Chicago Egyptologist John A. Wilson was struck by the last paragraph of the legend—Khonsu the smiter did not give up the whole of the tribute. Why? Says Wilson, this was a business commission, a payment he took for the trouble. Wilson may be right, but to me the statement has a mathematical ring. What the scribe is saying in effect is "Beware, these figures are an approximation." It may refer to the tolling of the sidereal months. On the tablet these intervals are precise only to within a day. But for 1270 B.C. that was a pretty good determination, particularly since the moon under perturbations is not truly regular, sometimes a day early, sometimes a day late. Another theory is that the approximation might refer to the beginning and end dates. The starting date is fixed by a solar festival, the feast of the valley, and the calendric length of the story of 6,447 days falls short of the full 18.614 years. "He did not deliver the whole of the tribute into the house of Khonsu" means it was short by 351 days, which is exactly 12 lunations.

The beginning and middle dates refer to festivals. In ancient Egypt these feasts went on for several weeks with tributes, banquets and all the pharaonic pomp and splendor. Are those dates pinpointed by the scribe especially important? Has he scored astronomical hits? Yes! The starting date is a full moon, the 1st full moon after the vernal equinox. The middle date in 1278 B.C. has no day number given, but it falls within the eclipse danger period for that year. So far as the computer can extrapolate (and here I rely on data kindly provided by Dr. Fred Espenak of NASA), there was a total eclipse of the moon on February 27, 1278 B.C., visible in all its phases from Thebes. Perhaps that is why that lunation was noted by the scribe—the princess was possessed by Humbaba! The closing date of the story has the moon crossing the ecliptic on its downward path near the descending node, the place where one would finish the count of draconic months, another appropriate spot astronomically!

Years before, when John B. White came into my Smithsonian office and urged me to publish the Stonehenge discovery, I was cautious. I played the devil's advocate and put the theory to the test. Were the errors too large, would a change in the age estimates by the archaeologists destroy the correlations, could earth movement, continental drift, affect the results? The answer was a 3-fold "no." Those errors were commensurate with a Stone Age culture, the sun and moon are not affected by the time change of precession, and continental drift would deviate the lines by only the smallest fraction of a degree. There was no weakness in the theory in those areas. Future criticism could be met.

With the princess legend I once again looked for a weakness that could be an Achilles heel. What if the latest dating of Rameses II is proved wrong? If the date of his reign is revised from the latest estimate, 1304 B.C. (a figure reached after more than 100 years of study by Egyptologists), this would not destroy the internal consistencies in the story. The journeys and the stay are fixed intervals agreeing with the astronomy and are unaffected by the actual Rameses dates. Also the moon would be in the same constellation at each of the 3 dates given, though the star would not necessarily be Antares. I asked, could this all be due to random chance—the moon in the windows at the right year, the 18.6-year cycle showing in the dates, the waning princess associated with a moon-god, the wandering moon returning to precisely the same star, the scribe warning about an "approximation"—could this all be coincidence? Hardly. Already there were the astro-archaeological alignments in the temples marking the high and low moon extremes. These had been found independently, and now this hieroglyphic account fitted that moon movement. And the tablet was found at the very site of these temples. It is the Stonehenge situation plus much more. In the Egyptian alignments we have confirmation in writing. Could the tablet be a pious forgery as claimed by Erman? No, the information is so detailed, so intricate, so correct for the time of Rameses II, that it can only refer to observations dating back to that ancient time. The tablet was probably copied from a papyrus script word for word. The scribe could not have calculated backward to make this fit because the Egyptians did not keep a running count of the years; time stopped and began again from year 1 with each new pharaoh. Finally I questioned the reigns. The temples of Mut and Amon-Ra were there at the time of Rameses II, but the temple of Khonsu was not; it was commenced by Rameses III. Here, as at Stonehenge, there must have been a great deal of preliminary observation and staking out. The main Khonsu temple with its roof window is more a ritual alignment with a known moon direction than a pristine observational marker. It seemed to me, as I played devil's advocate, that the theory would ultimately survive these criticisms if they were raised.

At the end of this first look into mindstep 1, I began to see evidence for those special legends of journeys and numbers being astronomical. It was the ancients' way of recording the basic facts of their cosmos. Gilgamesh, Quetzalcoatl, and the waning princess are samples. There are others—the story of the Golden Fleece, Isis and Osiris, the Mesopotamian creation epic—but these require more work and astronomer-mythologist cooperation to decipher.

Numbers have always been a stumbling block to understanding—prehistorian Aubrey Burl admitted at a Smithsonian lecture that equations "terrified" him—yet they are the stuff the cosmos is made of. Obtaining them in prehistory gave the savants a grasp of the order in the greater universe. The gods on the stage with the starry backdrop were no longer to be feared as entirely independent things or beings. Their actions were somehow controlled or ordained by unseen magical powers. Yet the performance up there was not entirely assured. Blood sacrifice was needed in the Aztec and Mayan cultures to keep Venus on course, and tribute had to be poured into the Khonsu temple to propel the moon around its orbit.

But over the centuries in ancient Egypt the giant leap of mindstep 1 faltered and stood still. It became a secure and fixed mind-block, or mind-set. For thousands of years the sky was looked upon as a revolving panorama, a drama which the gods acted out year by year, occasionally coming down to intercede on earth. Everything was revitalized in cycles. The pharaoh was part of this process. For Egypt the next step might never have come. It was to require a jolt to their comfortable way of thinking, an infusion from outside. Yet in this mindstep story from the time of the pharaohs we have a glimpse of all those earthly pleasures that king Gilgamesh a millenium before had been commanded by the gods to seek. For Rameses there was gold, lapis lazuli and men with sweet wood on their backs, a queen who pleased his heart beyond anything else, a festival with many offerings, and a people who were joyful and happy.

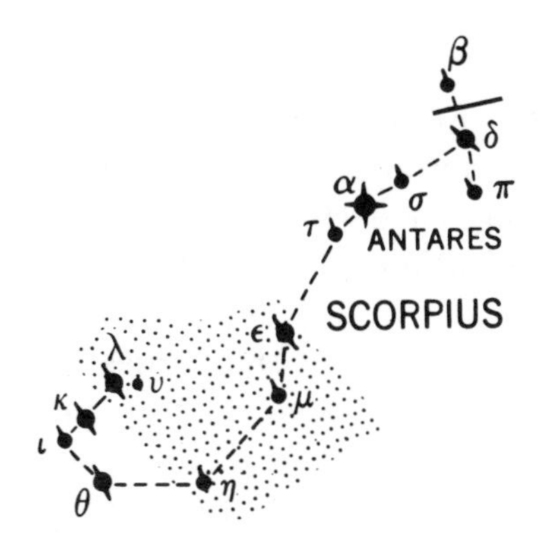

SEVEN
Mindstep 2

Something happened in those states which were to become classical Greece. A spirit of inquiry developed. Questions were asked. Philosophies spread. The cosmos was about to change from something "up there" to something "out there." The arena of the gods was to dissolve into a concept of space and time, the planets to become objects, not demons. The fabric of the heavens was about to be pulled aside. Why a new mindstep should have started in Greece we do not know. Perhaps it was because democracy began there, and freedom of the person led to freedom of thought and release from superstition. Perhaps it was because the Greeks invented the first alphabet with vowels—those 24 Greek letters that in combination could express all spoken sounds, ideas and thoughts. (Our English alphabet of 26 letters is named in tribute after the first two letters of the Greek—alpha, beta.) The Egyptian hieroglyphs, even though there were hundreds of them, had ambiguities and were never precise in the conveying of meaning. Or perhaps the mindstep began in Greece because of the inspiration of those lovely green islands under crystal-clear skies.

Thales of Miletus is one of the earliest philosophers we recognize, but we are not sure how much of his work was original, nor how much has been ascribed to him generously by later authors. Thales is credited with the prediction of an eclipse of the sun in the year 585 B.C., but if he did so he was lucky, because he had no theory to go on. Modern calculations confirm that there was indeed an eclipse (May 28 by the Old Style Julian calendar), and he probably observed it. He was about 55 years old at the time, in his scientific, philosophic prime. He also measured the angular diameter of the sun as 1/60 of the length of a zodiac sign, or 1/2 degree. The king praised him for this and proffered a reward, but Thales refused to accept, saying the prestige of the discovery was enough.

Anaximander, another Greek philosopher, argued about the nature of the celestial sphere. He said it was solid but pierced with holes through which the eternal fire could be seen, and that gave us the stars, sun and moon. Because the sun and moon move, their apertures must be in bands that slip along the sphere, and eclipses occur when the hole is covered up. He was wrong, of course, but he was asking questions and probing.

Pythagoras had another theory. It is difficult to decide how much of it was really his; the Pythagoreans formed what amounted to a secret society with closed discussions, being careful not to spread their ideas to people who were not members of the cult. One, Philolaus, was supposed to have broken the silence by writing up an account of the Phythagorean doctrine, but his original manuscript has never been found. Some historians even suggest that Philolaus was a straw man set up by Plato in his rebuttal of the theory.

In their system of the cosmos, the Pythagoreans had everything, including the sun, revolve around a central fire. They had broken away from the flat earth concept, which was quite a step forward for 500 B.C. The stars are hung on a sphere, they argued, so therefore the earth itself must also be a sphere. In their cosmos the celestial sphere was fixed, but the earth traveled around a circular orbit once every 24 hours. People in Greece faced outward to the starry sky, and people on the back of the earth faced the central fire. The panorama of the rising and setting of the sun and constellations took place because of the daily orbiting of the earth. There was a very slow rotation of the celestial sphere on crystal bearings. As it turned it generated the music of the spheres, which could not be heard by mortals (except, it was rumored, by Pythagoras himself).

But it was still too early for such ideas as a moving earth to take hold. Pythagoras was ridiculed and persecuted. Others said

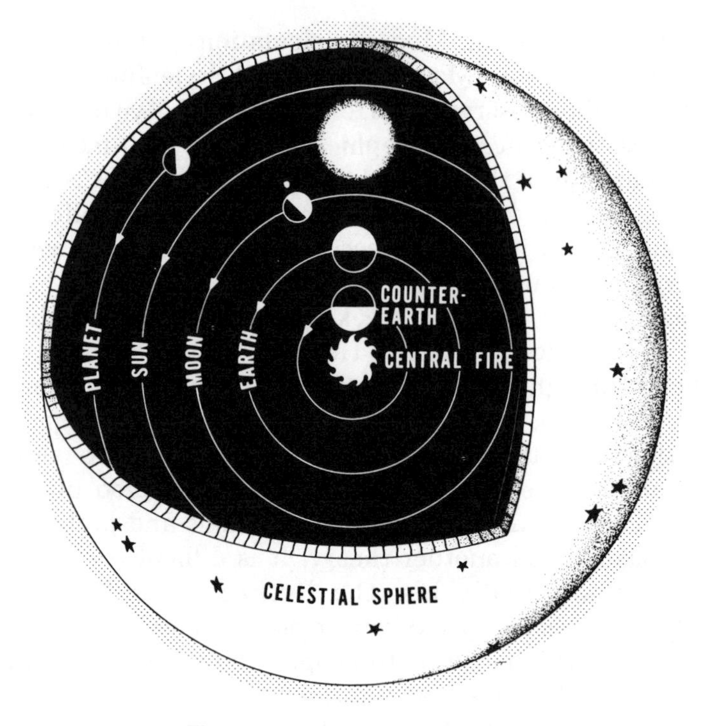

The central fire of Pythagorus.

it was an insult to the gods to imagine the base earth to be in motion like the sun and moon. "Also," said the enemies of Pythagoras, "people in India, on the other side of the earth, should see the glowing, eternal fire, but they do not." Undaunted, the Pythagoreans proposed a counter-earth, a sinister alter ego to the earth moving around the central fire once every 24 hours, to shade the people in India from its glare. "Why do we not see the counter-earth? . . . Why does it not blot out the stars?" The answers to these awkward questions were known only to the cult's inner members, so they said.

Advancement of knowledge was slow because learning was reserved for the privileged upper classes; because communication was difficult; because books were rare and expensive; because travel was dangerous; and above all because the collective mind was still paralyzed by those long millennia before philosophers had dared to ask questions.

Alexander the Great unknowingly gave an impetus to the new mindstep. Not that he was in the group of famous thinkers, writers and philosophers who led the way. Far from it. To establish a military empire by the age of 25 was fame enough. He was very much a straight product of mindstep 1, seeking proof of the nodding of the priest's head in Siwah, Egypt, that he was truly the earthly son of celestial Amon (Zeus to the Greeks). But by his conquering of Egypt, Greek thought and ideas swept into that ancient land behind his sword. After Alexander, Ptolemy ruled Egypt. He founded the Greek colony of Ptolemais near modern Benghazi and, most important of all, started the library of Alexandria, which was to become the core-memory of all ancient intellect and knowledge. This library grew in content, wisdom and fame for more than 600 years, fed by royal patronage.

Unfortunately nothing remains there today. We do not know how many irreplaceable books were destroyed when the mob burned it down in 389 A.D. Some claim that the library had already fallen into disrepair, that many books had been lost or stolen, and the fire finished the destruction. Others see in the ashes of Alexandria an estrangement from the past that will never be overcome. One way or the other, the books of ancient philosophers have gone, and their ideas on science come to us at second hand and by hearsay. Judging from the references in a few existing relics, at least 90 percent of the ancient books are gone, and we shall never know their true content.

Eratosthenes was the head librarian around 200 B.C. He agreed with the Pythagoreans about the spherical earth, and he set about measuring its size. On midsummer's day there were no shadows at noon at Aswan in southern Egypt, and the reflection of the sun could be seen in the water of a deep well. This meant the sun was exactly overhead in the zenith. But at noon on midsummer's day in Alexandria, the sun was a little to the south of the zenith. The buildings cast a small shadow the length of which showed how the sun was displaced from the zenith by $\frac{1}{50}$ of the circumference of the heavens. Eratosthenes paid men to pace out the distance between Alexandria and Aswan, and they found it was 489 miles. The distance around the earth must therefore be 50 times as great, Eratosthenes said, or 24,450 miles, making the diameter 7,800 miles. Not a bad estimate for 200 B.C.!

Hipparchus, on the island of Rhodes, became the first obser-

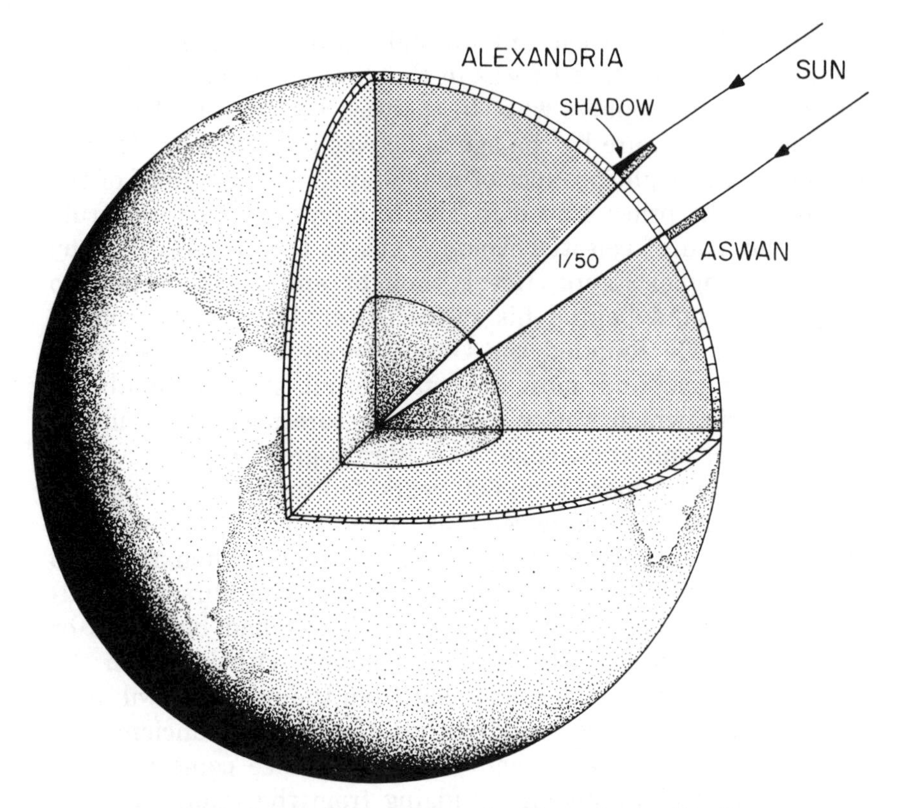

Eratosthenes's measurement of the spherical Earth.

vational astronomer—or at least the first whose name has survived. He carefully prepared charts of the sky, showing the position and brightness of each star. By night he followed the motion of the moon and planets among the stars of the zodiac; by day he measured the altitude of the sun from the length of the shadow of an obelisk. From his measurements he discovered the drift of the equinoxes from Taurus to Aries. Since his time, the vernal equinox has continued its movement and is now in the middle of the constellation Pisces. Hipparchus also found that the sun did not move at the same speed around the zodiac each month, moving faster in January than in July, owing to the (then unsuspected) elliptical orbit of the earth.

It was an astronomer named Ptolemy who was going to make the next step. Unlike Hipparchus, he was primarily a theoretical astronomer, not observational. We know nothing of his personal life—only his books, his discoveries and his system of the cosmos. He was born in Ptolemais and took his name from the town, and hence, with a 500-year time lag, from Alexander's ex-bodyguard, Ptolemy, who had taken the two crowns of Upper and Lower Egypt as Ptolemy I. We know Ptolemy the astronomer was alive in 127 A.D. because of a measurement of a planetary position which can be traced to that date. Living and working in the library at Alexandria, Ptolemy had the collected knowledge of the world at his fingertips. We do not know whether he was Egyptian, Greek or Roman—his Latinized name was Claudius Ptolemaus—nor do we know if he was a triple genesplice, but certainly he was influenced by all three cultures.

In Ptolemy's new system, the earth was at the center and the sun, moon and planets were supposed to revolve around it at spaced distances, the moon being the nearest and Saturn the farthest away. Uranus, Neptune and Pluto, of course, were yet to be discovered, and comets and meteors were dismissed as nonastronomical lights at the top of the atmosphere. The stars were hung on the inside of a revolving, crystalline sphere.

The planets (and sun and moon were included in that category) moved each on a circular epicycle, and the center of the epicycle moved in a perfect circle around the immobile earth. Mercury and Venus were exceptions—those inner planets moved around their respective epicycles, which were centered on an imaginary arm between the earth and the sun. A planet traveled once around the epicycle track in the length of the synodic period (program AGE). For the outer planets the center of the epicycle traveled around the zodiac once per sidereal period. Complicated? Yes! But Ptolemy was forced to this system if the earth was to be still. He needed to invent the epicycle to explain retrograde motion—that time in the cycle when a planet seems to go into reverse for a few weeks, at opposition or inferior conjunction (program AGE).

It was a complicated system, yet with one creative step, a mortal had brought the planets under mathematical law and created a new concept of space and time and things—at least for that region of outer space containing the sun, moon and planets. Motion in a perfect circle had long been accepted by the philosophers as something akin to perpetual motion, and by adding cir-

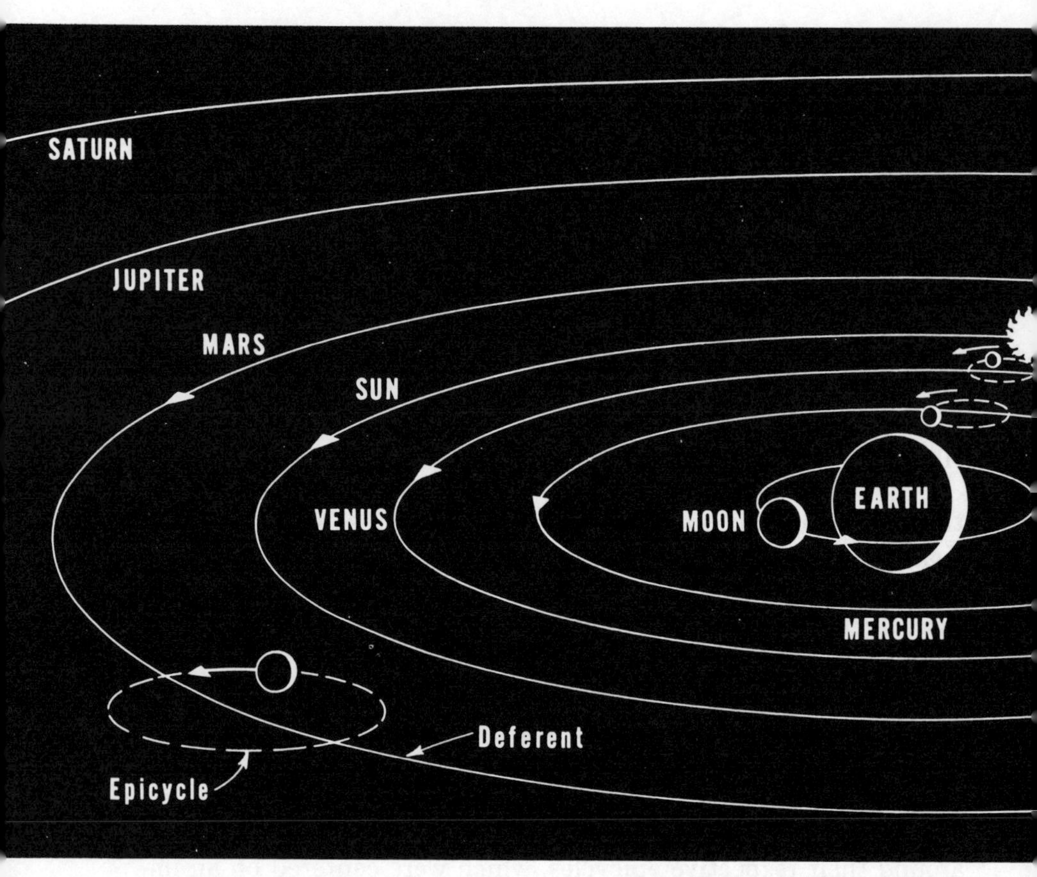

The Earth-centered cosmos of Ptolemy.

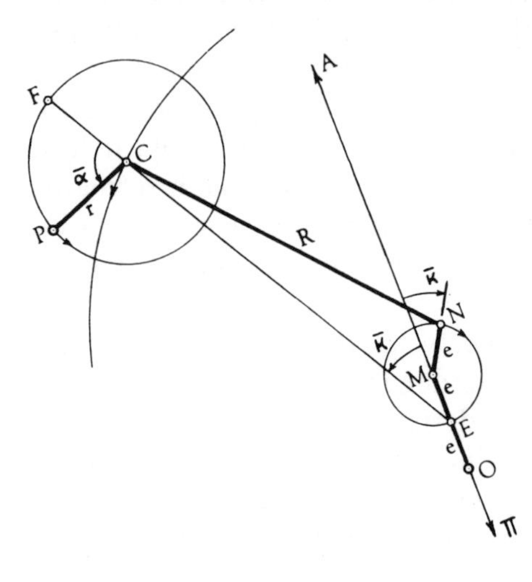

Ptolemy's geometry for explaining the motion of one of the planets: Mercury.

cles Ptolemy did not need to invoke physical forces. The magical forces of the cosmos would suffice. He would be spared the awkward, practical questions asked of Pythagoras.

Ptolemy published mathematical tracts, wrote a treatise on global geography which was the standard reference for a thousand years, inspiring Columbus to sail west to get to the east, and he worked unceasingly on the perfection of what was to bear his name as the Ptolemaic system. Mathematically it was indeed perfect. All celestial movements, all perturbations, were accounted for by adding extra centers, extra circles, almost ad infinitum.

To some the pocket-calculator programs in the Appendix may look fearsome, but, effectively, they contain only the first terms (the circles) of the equations. Ptolemy and his successors added on hundreds of terms. And they worked with pen and tables of numbers, not pocket calculators! The more work one is willing to do (for us, the more buttons we are willing to punch), the more accurate the result will be. In principle his system would be adequate for programming those precise calculations done by the high-powered digital computers of the space program. It could track the movement of an orbiting space colony, or trajectories in star wars. Take the relatively simple problem of predicting how the sun will speed up and slow down between summer and northern winter. We know this is due to the elliptical shape of the earth's orbit. Ptolemy took care of the problem in the following way: instead of the sun's deferent (the track of the center of the epicycle) being drawn exactly around the earth, he put it a little bit to one side. This adjustment would cause the sun to seem to travel faster when on the near portion of the deferent, and more slowly when on the far portion. With his full system, and with the periodicities already established during the long millennia of mindstep 1 before him, Ptolemy was able to predict the movement of celestial objects as accurately as was needed.

If Ptolemy had been a politician or potentate he could not have done better. The Ptolemaic system was practical, satisfying and comfortable. The planets could no longer be gods if they followed the scribblings of a pen, yet there was room enough for the transcendental in the space beyond the crystalline sphere. The earth was at the center of the universe as everyone expected it to be. Earth, air, fire and water were separated. Heavenly things moved in circles, earthly things stayed put. There was a difference between the shining lights in space and the solid environment of earthlings. Earth did not shine, nor was earth a planet. Earth was not balanced on the backs of 4 turtles in a pool of water; it was fixed, nonrotating, immovable, at the place where no disturbance could ever affect it—at the very heart of the cosmos.

With Ptolemy's work completed, his scrolls written out, edited and placed in the library at Alexandria, the world began to move forward in mindstep 2. It moved with the inertia of the Ages, the weight of religious and sociological change, but once the triggering impulse had been clearly applied there could be no turning back.

EIGHT
In the East

A mindstep takes hold slowly, but relentlessly. Ideas pass from brain to brain, changing the whole thought pattern. A dominant gene shapes the body of future generations, and a dominant meme shapes the mind. If it is not dominant it ultimately perishes, swamped by competing ideas. A mindstep once established will last indefinitely, like a species of plant or insect, until the next mindstep comes along. Just as a species requires a favorable environment to propagate, a mindstep takes hold and spreads when the conditions are right.

Of course, it is difficult to place a date on the early mindsteps. The first look at the cosmos, the sense of wonder of mindstep 0, probably occurred around 35,000 B.C. when people began to draw pictures and carve figurines. It coincided with the invention of visual representation. It may have taken place much earlier when communication was limited to speech, but that would be difficult to prove.

I place mindstep 1 sometime between 7000 and 3000 B.C.—probably nearer to the later date, coinciding with the invention of writing. Again it is entirely possible that movements in the cosmos were explained by stories much earlier than those dates, but that is difficult to prove on the basis of the record. The myths and legends seem to be unfathomably old, but the first clear proof of the obsession with a pantheon of gods overhead comes with cuneiform writing and hieroglyphic texts.

It is interesting to speculate what would have happened if the earth had been a cloud-shrouded planet. Sapiens the Intelligent would not have seen the stars. Quite possibly those early ancestors of ours would have continued to pick berries and hunt. They would never even have reached the starting point of mindstep 0. They would have become an entirely earth-oriented species.

On the other hand, they might have developed philosophy, religion and technologies underneath the clouds. Philosopher A might have speculated that there was something above the clouds. Philosopher B might have concluded that the earth as a globe must be moving through space. In that scenario we can imagine a culture getting to physics and the rocket in pure cosmic ignorance, and when the telemetry equipment broke through the clouds discovering the universe all of a sudden. Then, if they had the mind to absorb it, they would go from mindstep 0 to the end of the line all at once. . . . Perhaps, but somehow I doubt it. I think vision is essential, and the process of adjusting to what is seen is a slow one. Sapiens, I believe, need inspiration as well as knowledge.

Mindstep 2 is dated to the lifetime of astronomer Ptolemy. He died in approximately 150 A.D., so one can use that date or, arbitrarily, the year in which he made an astronomical observation, 127 A.D. There had been precursors for this mindstep. Aristarchus of Samos said the earth was spinning on an axis and moving around the sun in an orbit. This was as far back as the year 270 B.C. He was correct of course. What he was saying encompassed mindstep 2 and mindstep 3, but he was too early. The collective mind was not ready for the double step and his work was totally ignored at that time. Like the moon paintings in the Stone Age caves, it was an isolated precursor. In fact, Aristarchus was opposing the words of the great Aristotle, who was more eloquent, more logically convincing, and whose reputation was to go unchallenged for centuries. He had argued for a fixed, immovable earth. For all of prehistory the earth had been terra firma, and the ancient world of Aristarchus was not ready for the truth. Aristotle won though, and Ptolemy accepted his viewpoint. Earth below was to be fixed and immovable and separated from the cosmos "out there." But Ptolemy adopted one idea of Aristarchus's: the earth was a perfect sphere. Never again would it be the flat disc of mindstep 1.

If mindstep 0 depended on drawing and mindstep 1 depended on writing, then mindstep 2 depended on mathematics. Each of these was an aid to thinking and the broadcasting of an idea. They could be classed under the broad heading of communication technologies, because even mathematics is in some ways a language and creates a transmittable record.

Ptolemy benefited from the geometrical theorems of Euclid, and he himself developed refinements in the calculations of angles, but the full force of the Ptolemaic system required mathematics that was yet to come. Planets were no longer spirits of the astro-gods, they were simple points of light whose movement could be predicted—Unidentified (but predictable) Flying Objects. The power of prediction stimulated the steady growth of mathematics throughout the long period of mindstep 2. The cosmos marched on numbers and the earth stood still. There was now a rational explanation for those complicated motions in the zodiac, and behind the wheeling perfect circles there was room for a master designer. Ptolemy's system fitted harmoniously with one-god religions. Mindstep 0 was the Alpha Age, mindstep 1 was the Age of Myth and Legend, and mindstep 2 brought in the Age of Order.

Eclipses were now understood and could be predicted mathematically. Terror was replaced with wonderment. The retrograding planets conformed to the rules of the epicycles. The zodiac was the beginning and end of the cosmos.

The zodiac, of course, needed to be made into a wide band to contain the tilted paths of the moon and planets. The sun's track cuts through the middle of the zodiac band exactly. Only when the moon crosses this imaginary sun-track can an eclipse of the sun or moon take place—hence the Greek-derived name "ecliptic." At the time of crossing, the sun and moon are directly in line with the earth in space. When the moon is in front of the sun, the sun is blotted out. When the moon is opposite the sun, the moon is blotted out. The earth's shadow reaches out to the orbit of the moon and makes a dark patch in space. Normally it can't be seen because it is in the night sky, but when the moon is full and exactly on the ecliptic track it falls into the earth's shadow. Instead of shining in the bright sunlight of deep space the moon darkens to a dull, blood-red coppery color. The only sunlight to reach it has been filtered through the dust and haze of the earth's atmosphere. When the moon is new, on the other side

Eclipse of the moon.

of its orbit, the tip of its cone-shaped shadow touches the earth. Earthlings in the center of the cone see the black silhouette of the moon covering up the sun. Those at the edge see a partial eclipse; a portion of the sun's disc is eaten away. An eclipse of the moon can be seen by half of the earth. An eclipse of the sun is more difficult to see because the moon's shadow is so small and touches a particular place on earth only once every 200 or 300 years, but sun and moon eclipses were a predictable quantity in the Ptolemaic system.

Inner planets Mercury and Venus were supposed to go around their epicycles with their own synodic periods while the

Eclipse of the sun.

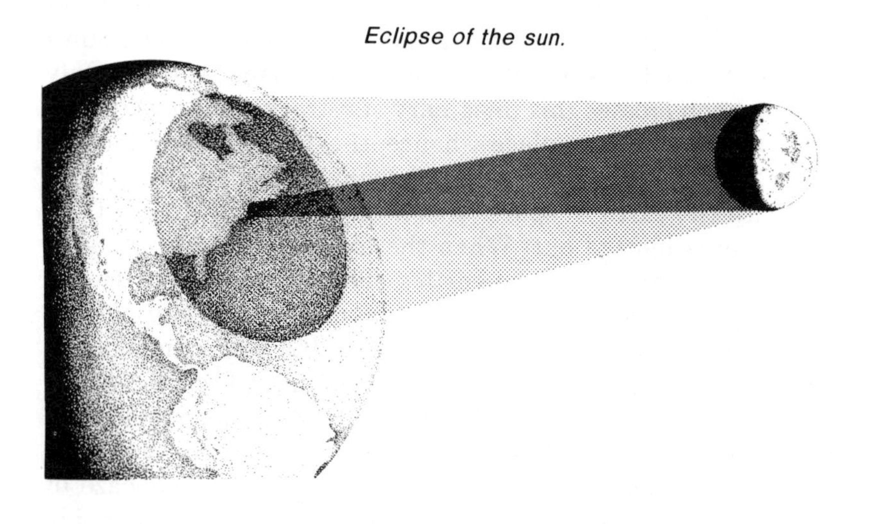

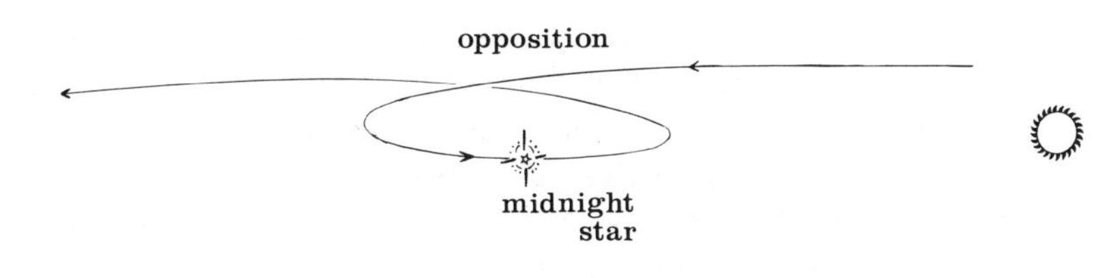

Retrograde motion.

arm to the sun carried them around the earth—a convenient computation model, it is used in program AGE.

Outer planets Mars, Jupiter and Saturn make a great sweep around the circle of the zodiac. The planet first seems to emerge from behind the sun as a morning star, then week by week the separation from the sun (called elongation) increases. Halfway through its cycle the outer planet moves in the sign of the zodiac directly opposite to the sun. If the sun is in Aries, then the planet is in Libra, if Taurus, then Scorpio, and so forth. At that time the planet would be eclipsed like the moon if the earth's shadow were long enough, but the outer planets are far away and the shadow cone cannot reach them.

At opposition a planet rises at sunset and is over the due-south line at midnight. It reaches maximum brilliance at that time because it is closest in orbit to the earth. The planet now retrogrades against the stars: instead of moving slowly eastward in the zodiac it appears to go westward, reversing the normal order of signs. The normal order is Aries, Taurus, Gemini . . . but the retrograding planet goes Gemini, Taurus, Aries. This is because the earth, moving in its own smaller orbit, is overtaking the outer planet. After retrograding for a few weeks the planet resumes its journey around the zodiac until it finally disappears in the west as an evening star. From there it goes to superior conjunction behind the sun. It is moving so slowly that the sun appears to pass in front of it. The planet shows up as a morning star and another cycle commences. These aspects can be figured on a calculator with program AGE. Fierce though AGE may look to the mathephobe, it is a simplified routine ignoring the epicycle. Ptolemy would have added the equant and epicycles to the program to improve the accuracy.

By its nature a mindstep eventually engulfs the whole earth. It springs up at one spot and spreads to the four corners. One of the first communication technologies—speech—probably started in different areas at different times. Mindstep 0 probably appeared in various cultures independently rather than diffusing across boundaries. The isolated Australian aborigine drew pictures that represented the cosmos, and so did the African tribes like the Dogon. Pre-Columbian America advanced through mindsteps 0 and 1 in the classical period, yet we have no evidence for those mindsteps traveling across the Atlantic from the Spanish caves or Babylonia. It seems the Maya and related cultures invented writing and developed the myths and legends of the astro-gods without the Indo-European influence. But mindstep 2 started in at a single point in space and time, in the great library of Alexandria, and we can trace the way the shot was heard around the world.

In the Middle East and in Europe the dome of heaven was rent open, the panorama of gods acting out their drama overhead dissolved away, the immortal demons that plagued Gilgamesh were bridled and forever after would follow orderly paths computed by man, ordained by God. The underworld beneath the flat earth of mindstep 1 became an inferno at the earth's center. There was a heaven-space, an earth-space and a hell. The Ptolemaic cosmos was in comfortable harmony with the great monotheistic religions—Judaism, Christianity, Islam.

On the Indian subcontinent people were receptive to the new mindstep. Hinduism already accepted an orderly universe, and space was regarded as infinite. Time proceeded in great cycles lasting 4.32 billion years. For the Hindus there was no fabric to be rent, no dome to be smashed, and Ptolemy's deferents and epicycles had room enough to be deployed in their conceptualization of infinite space and time. Indian mathematicians quickly adopted the epicycle as the explanation of retrograde motion and worked out algebraic methods for computing the position of the planets. Their interest was mainly astrological. The planetary positions were thought to control the destiny of rulers and even those of lower castes, and they took advantage of the new orderly, geometric planetary system. Previously the retrograde motion had been blamed on a spirit-power that lay on the orbit at a place opposite the sun. Now it was a mechanical, predictable swing.

In the pseudo-science of astrology in India and elsewhere the life of a person was thought to be influenced by the sign of the zodiac, where the sun was at the time of birth (or in some astrologies, the time of conception). A planet 120 degrees away from the sun was in "trine"—a good configuration. A planet 90 degrees away was in "quadrature"—bad. Retrograde motion was bad. Direct motion—Aries, Taurus, Gemini—was good. Conjunction of 2 planets could be propitious or malevolent depending on the pairing. Venus was beneficent, second only to the good fortune brought by Jupiter. Mercury was an influence on the critical faculties, bad when retrograding, and Saturn was almost always a bringer of misfortune and connected with death unless modified by the aspect of Venus or Jupiter. The moon and Mars also had their supposed effects, and all was controlled by the attributes of the actual zodiac signs and their positions with respect to the horizon and the due-south line. At this level astronomy and astrology were closely related, and observation and prediction were both used for horoscopes.

Astronomical though they are, the pocket-calculator programs are by the nature of the sky connection also horoscope aids—the manifestations, retrogradings and conjunctions with the sun can all be tapped out with ASTRODATE and AGE. For example, what about the "Star of Bethlehem" seen as a portent of the nativity of Christ? Some astrologers have computed it to be a simultaneous opposition of Jupiter and Saturn in 7 B.C., September 13 on the Old Style calendar. At that time the value of ASTRODATE was −5.3018, Jupiter's AGE was 200.0 days, and Saturn's 182.2. So Jupiter was at opposition and Saturn very nearly so. The planets were at their brightest, in close conjunction and retrograding. Actually the planets moved together in the retrograding loop for more than 2 months, a rare and (astrologically) notable event, coming together on 3 different nights—a triple conjunction. By modern astrological interpretive rules, Jupiter foretold the greater beneficence, the peace on earth, and Saturn foretold the crucifixion. Whether or not the Magi would have interpreted the portent in that way we do not know, nor is the date of the Nativity historically confirmed, but this conjunction is the most frequently suggested astrological occurrence. I mention it here as an example of the type of mathematics one deals with at the common root from which astrology and astronomy have diverged.

Interestingly enough, the Indian astrologers knew that the Ptolemaic system was not perfect. They improved it by replacing the circle of the deferent with an ellipse, and they rejected the idea of the offset equant. While the Europeans added epicycles to epicycles to improve accuracy of prediction, the Hindus did not. They left the mathematical discrepancies to be explained away as due to "winds," or "unknown forces." Yet the mindstep as it spread to India was beneficial, leading to the invention of the concept of zero, the development of spherical trigonometry (triangles drawn on a sphere), a better value for π ($3^{177}/_{1250}$) and a precise astrology. Accurate observations were made of the sun, moon, stars and planets, and at the end of the era we find the Maharaja Jai Singhe building enormous brick sundials and measuring quadrants in the then new city of Jaipur, 100 miles southwest of Delhi.

Ancient China was resistant to mindstep 2. Chinese scholars were fully capable of adopting the idea of orderly planetary motion and computing positions far into the future, but they rejected it. Explorers like Marco Polo, traders and missionaries exposed the East to Western thought, but China was not ready for the step. There was a counter-meme which would hold ancient China in its grip almost to the modern era.

China was humanistic, artistic, philosophic and mystic. These are the thought patterns of the right hemisphere of the brain. The deductive faculties of the left hemisphere were suppressed. As late as 1050 A.D. scientist-philosopher Shao Yung declared: "In science we must never attempt to force things we do not understand into a scheme, into a system."

The Chinese saw the cosmos as an organism with separate parts. It was not an ordered universe, not a universe governed by an unseen controlling force. For them, the sun, moon and stars moved through a harmony of wills, like a troupe of coordinated dancers. There was a measured pattern, yet there was always the possibility of something, someone, breaking up the harmony.

Ancient China went as far as any culture could go without heeding the new mindstep. Chinese observational astronomy was far ahead of the rest of the world; their annals contain details of meteor showers, eclipses, nova and supernova stars. Earth sciences—surveying, mining, prospecting—flourished, as did medical sciences—acupuncture, smallpox immunization as early as

1000 A.D. In technology they invented gunpowder, rocket-launched arrows, the magnetic compass, mechanical clocks, paper currency, enamels, lacquers, plastics, and the movable-type printing press. But in Cathay fundamental knowledge was looked for in the mystical *Book of Changes,* the *I Ching,* instead of in the physical world, and without a search for order in the heavens China was destined to lag behind those events unfolding in the West.

Sometime before 600 A.D. a large iron meteorite looped toward the earth from beyond the orbit of Mars and smashed into Arabia. A flash, a heat-pulse, a pillar of fine dust, sand was turned into glass, and then silence. This was in Ar-Raba-Al-Khali, the so-called Empty Quarter of Saudi Arabia, the lifeless and life-destroying desert. The iron hulk of the visitor from outer space lay in this lonely place to be covered and uncovered and covered again by the endless wind-blown sand. Today it has become a spot on the empty map, Al Hadidah, the "place of iron."

This desert has the harshest climate on earth. Each year it bakes in a record-breaking 4,000 hours of sunshine; the ozone layer overhead is at the earth's thinnest, so the ultraviolet light is extreme; rainfall is less than 3 inches per year, essentially unmeasurable; at high noon with the sun directly in the zenith, temperatures (in the shade) reach 140° F. On those very rare occasions when it does rain, locust eggs hatch out and genetically coded insects fly off as a plague to the nearest green leaves in the Nile valley or Central Africa.

Harsh climates produce tough people. Arabia was the original home of the ancient Semites, who spread east to Babylonia, north to Syria and west to the borders of Egypt. Those tribes lived on the fringe of the desert, tapping underground water left by the melt of the Ice Age thousands of years before. Travel over the desert was accomplished by camel caravan, usually at night; navigation was by the stars.

Meteorites are the left-over fragments of a small malformed planet between Mars and Jupiter. The planets of the solar system collected from the dust and gas surrounding the proto-sun some 5 billion years ago. But something went wrong between Mars and Jupiter. Either the temperature of the dust cloud was not conducive to growth, or the gravitational field of giant-planet

A slice from an iron meteorite, Meteor Crater, Arizona, showing large crystals.

Jupiter prevented the coming together of the material, and a few small moon-size objects formed instead. Occasionally the fragments land on earth, out of the sky. They are made of stone or iron, or a mixture of both. The so-called "stones" contain round chondrules, ball-bearing-size crystals which may ultimately prove to be the oldest minerals in our solar system. The irons are large single crystals, testifying to a cooling process lasting a thousand years or more in space. Irons and stones show evidence of fragmentation, probably caused by collisions in space.

As well as being the home of the influential Semites, Arabia was and is the focal point of the Muslim religion, Islam. The most important shrine is the holy Kaaba (Ka-aba) in Mecca. The name translates as "cube," and that is the approximate shape, between 30 and 35 feet on each side. Dating from the time of Ptolemy, or by tradition to the time of Adam, there was a pre-Islamic shrine at Mecca connected with earth- and sky-gods. Tradition says that Abraham rebuilt the shrine—the stone on which he stood has been preserved in a special chapel to the northeast of the Kaaba.

There is a meteorite set in a silver yoke at the southeast corner of the Kaaba. Patriarch Abraham, tradition says, was brought the black stone from heaven by the angel Gabriel. When pilgrims make the *tawaf,* a counterclockwise walk around the Kaaba, they start and finish at the black stone. They make 7 revolutions, the first 3 quickly, the last 4 more slowly. Some of the traditions say the tawaf ritual represents the cosmos: 3 circuits for the fast-moving moon, Mercury and Venus, and 4 for the sun and outer planets Mars, Jupiter and Saturn. Other traditions say the tawaf represents the circulation of angels around God in highest heaven.

The Kaaba meteorite is thought to be of stone, but no chemical analysis has been made to determine its composition. All non-Islamic persons are forbidden in the holy city. Those infidels who attempted to break the law in the past were beheaded with a golden sword. Tradition says the stone of Mecca was white when it first came down and then later changed to black. This would be consistent with the oxidation process of a stoney iron. Most stone meteorites contain flecks of a nickel-iron alloy and these inclusions can color the stone.

As an astronomer I wondered if the Kaaba was aligned to the sun, moon, planets or stars. Writing in Arabia was invented at the time of the Prophet, around 600 A.D., the holy Koran of the Muslim faith being the first book ever to be written in Arabic. Before then words were recorded in the memory, written on the heart, carried in oral tradition. By coincidence, David King of New York University, an expert in Islamic traditions, also wondered the same thing, and he got in touch with me. He showed me an ancient manuscript from 1290 A.D. which referred to alignments. It was written by Muhammed ibn Abi Bakr al-Farisi, and Professor King translated as follows:

> The sun rises on Kanun I, the 19th, which is the shortest day of the year and the beginning of winter . . . and sets between the south corner and west corner of the Kaaba, which is where the lunar crescent first appears in that month. . . .
>
> The first of the winds is the saba which is easterly of the Kaaba facing the black stone and the southern corner, and it blows from between sunrise and the rising point of Canopus.

Al-Farisi gives the alignments along the perpendiculars to the faces of the walls, not the line of the foundation. I took "facing the black stone and the southern corner" to be the long axis of the shrine (its base is not exactly square and is elongated in that direction). "Between the south corner and the west corner" seemed to be the shorter axis, which is almost, but not quite, perpendicular to the long axis.

Using the STONEHENGE program (see the Appendix) and the data obtained from the municipal map (neither I nor David King were or would be qualified to visit the site to make measurements), we computed moon and star alignments:

	Azimuth (degrees E of N)	*Skyline (degrees E of N)*	*Declination on Skyline*	*Target Indicated by the Traditions*
Minor axis	236.4	4.4	−28.7	First crescent of the winter moon, setting
Major axis	147.8	3.2	−50.2	Canopus (alpha Carinae), rising

The traditions were confirmed. The declination of Canopus around 0 A.D. was −52.6, and the declination of the center of the disc of the moon with the bottom cusp of the crescent just touching the Mecca skyline was −28.9. The star alignment was accurate to 2.4 degrees, the moon crescent accurate to 0.2 degrees. The latter was better than the estimated uncertainty of 0.5 degrees due to not being able to make measurements at the site.

The architectural right angle built into the structure had picked out the southern moon extreme of Callanish and other ancient sites, and the second brightest star in the sky, Canopus. The right angle of the Khonsu temple on the bank of the Nile also paired the moon and Canopus, though in Egypt the targets were summer moonset, not winter, and Canopus setting, not rising.

Professor King could find no explanation for this Canopus-moon pattern in the Islamic traditions, nor did the legend of the waning princess explain the Canopus connection in Egypt. But the Islamic traditions are full of references to the crescent moon. The crescent and the star are on the flags of Islamic countries, and the Muslim calender is strictly regulated by the moon. Whether or not the star on the flag is Canopus we do not know.

Al-Farisi's 13th-century text mentioning Canopus and the crescent moon.

Some say it is Mars, others say it is Venus morning star. But whatever the star connection, a person in pre-Islamic times looking along the line of the wall from the black meteorite would see the crescent in the west at the extreme point of its 18.6-year cycle. A person standing at the sanctified stone of Abraham looking at the front face of the Kaaba would see the moon setting over the center of the shrine. The crescent would swing to the right winter by winter as it did for a person within the station-stone rectangle at Stonehenge, returning in its cycle to align again with the building's axis. In the reverse direction the line would pick out the early-morning last crescent, with the crescent facing to the left as in Islamic flags, though this alignment is not as accurate as the evening one.

Allah's prophet Muhammad received his holy calling in 610 A.D. on the Mountain of Light, a hill overlooking the Kaaba. It was on the Night of Power, which tradition says was on the 27th

19th-century photograph of the Kaaba covered with a black drape. The crescent moon at midwinter set over the distant hills on the line of the Kaaba.

day of the lunar month, at which time the moon was in the last-crescent phase. On that night the Koran, or the first parts of the holy book, came down to mortals on the earth through the prophet. Each year on the night of the 27th the heavens open for all Muslim believers, and on that night Allah, the most gracious and most merciful, hears all prayers.

Islam spread rapidly in the 7th and 8th centuries, through holy wars and conversions, into the Mediterranean countries, Africa and India. It has now grown to be one of the largest and most powerful religions of the world, second only in numbers to Christianity, with one-seventh of the world's population being Muslim.

Arab scientists with their pens followed the sword, and there was a flourishing of caliph-supported learning. Ptolemy's *Almagest* and other classical works were translated into Arabic and written down in the new script, and all areas of planetary and mathematical astronomy benefited. It is Islamic writers who gave us the words for algebra, azimuth, zenith, nadir, altitude, and even the title for Ptolemy's book: "Almagest" means "the great," and it was applied as a recognition of his stature as an astronomer and the impact of his work.

These sharp-brained astronomer-mathematicians opened whole new areas of study. They published hand-written trigonometric tables, added epicycles to the Ptolemaic system, calculated the times of appearance of the moon's first crescent in lieu of the less reliable visual observations for calendar control, and they developed the mathematics for prediction of eclipses. They improved the techniques of planetary observation and timekeeping, and developed new computational methods.

For a full 7 centuries there was a flood of Arabic manuscripts which expanded knowledge and influenced the Western world. It was a powerful climax in the mindstep, the Age of Order. Yet precursors of change were there. The Ptolemaic system, despite the later refinements, was not entirely satisfactory. As early as the 13th century, an Islamic writer, al-Tusi, challenged Ptolemy and the whole basis of the earth-centered system.

But the next mindstep revolution was not to come from the brilliant skies of the Mediterranean. It was to come from a place far to the north, the cold and misty plains of Poland.

NINE
Mindstep 3

Nicolaus Copernicus was a man of all seasons. Educated at the universities of Cracow, Bologna and Padua, he knew everything there was to know in Renaissance learning, from the Greek classics to medicine. He was given the position of Canon at Frauenburg cathedral by his mother's brother, Bishop Waczenrode, in 1497, but he postponed ecclesiastical duties for 6 years to study astronomy and get a doctorate in canon law. Ptolemy's Age of Order was now the decidua of dogma, and Copernicus was expected to believe without question that earthlings lived at the center of the universe, that everything else moved about them in perfect circles, and that the earth, nonplanet as it was taken to be, stood still.

Canon Copernicus worked at Frauenburg for more than 30 years on things ranging from the practicalities of monetary control and inflation to the finer points of pure theology. Astronomy was a life-long hobby. Trusting his own mind before the opinions of others, he was aware of the shortcomings of the Ptolemaic system. Finally, after years of pondering, he decided the movement of the planets could be described more satisfactorily if we took the sun as fixed and the earth as moving. With that step the earth took its place among the planets. The system of circular orbits based on the sun was much less complicated than Ptolemy's epicycles. The retrograde motion of an outer planet was simply a result of planet earth overtaking it. As earth moved by on the faster inner track, the outer planet appeared to lag behind. Thus Ptolemy's epicycle would be needed no more because it was invented ad hoc to explain this retrograde motion.

Copernicus argued that the sun was a source of light and energy, was larger than the earth and was not like the planets, and so it was more natural for the small earth to travel around this unique object. With the earth in orbit around the sun, Copernicus argued for a second motion, a spin on an axis. Then the rising and setting of the sun and stars would be caused by earth's rotation. Earthlings were specks on a tiny globe, turning in the rotisserie of sunlight. He argued that the spin could not be felt by earthlings as a wind because the air was carried around with them in the rotation. To check his ideas, Copernicus made observations and did calculations with the new equations of spherical trigonometry which had come from Hindu and Arab mathematicians.

Although Copernicus had created the 3rd mindstep, his work was almost lost. He feared for the impact the idea would have and kept it almost a secret, sharing it with only a few. For 30 years he weighed his words carefully as he wrote out his arguments page by page. He realized those pages if exposed might be destroyed as a heresy, but for him there was no way to escape the pressure of the mindstep. In his manuscript he pointed out that Ptolemy's theory was not the only one to be proposed by Greek philosophers, and that one more theory from a canon of the church might not be improper. At one point Copernicus had been asked to help in straightening out the errors in the calendar, a Vatican project which was to culminate in the Gregorian calendar reform of 1572, but he declined to assist on the grounds that our knowledge of celestial motions was not precise. However, he privately wrote in his manuscripts—which by now were book-length—that his work would give a better basis for predictions and aid calendar reform. Finally, as the book neared completion, he dedicated the whole work to the Pope in a beautifully written preface.

Copernicus had been theologically diplomatic; he could do no more. The sheaves of paper were left in his room in the tower at the cathedral to be disposed of in his will. However, George Rheticus, one of his former students, was dazzled by the new idea, and just before Copernicus died he took the pages to the nearest

Nicolaus Copernicus (from a woodcut).

printing press at Nuremberg and had the manuscript edited by Andrew Osiander (a Protestant), set in type and published as a book. Osiander gave it a title, *Revolution of Celestial Bodies,* and he added a second preface. Protestants at that time were careful of the dogma of the earth standing still, and Osiander said the idea of the earth moving was only a suggestion, a "hypothesis" for making the calculation of planetary motions more simple. This second preface was nothing more than an abject apology and was entirely out of character with the sweeping spirit of the rest of the book. Osiander did not sign his preface and for many years it was thought that Copernicus wrote it. But we now know that Copernicus would not have sided with the apology; he fully intended his book to start a revolution in thinking about our place in the cosmos. Fortunately though, Osiander in his meddling helped rather than hindered the mindstep because the double preface acted as foil and made it difficult to denounce the book as a heresy when it first appeared in 1543.

Ptolemy's mindstep was copied by hand on parchment, but Copernicus's was spread by the new printing press. By 1566 the number of copies in circulation had reached 900. Copernicus was speaking with 900 voices.

In 1616 the Roman Inquisition banned the book "until corrected," and sent censors far and wide to obliterate the words which threatened the dogma. We now know they got to fewer than 50 of the 500 library copies, and in England, Spain and Portugal, not a single copy suffered the black obliterating smear.

German and English astronomers calculated the positions of the planets using the new theory. Surprisingly they found the Copernican system no easier to compute than the Ptolemaic one,

The Copernican system makes the Earth a planet.

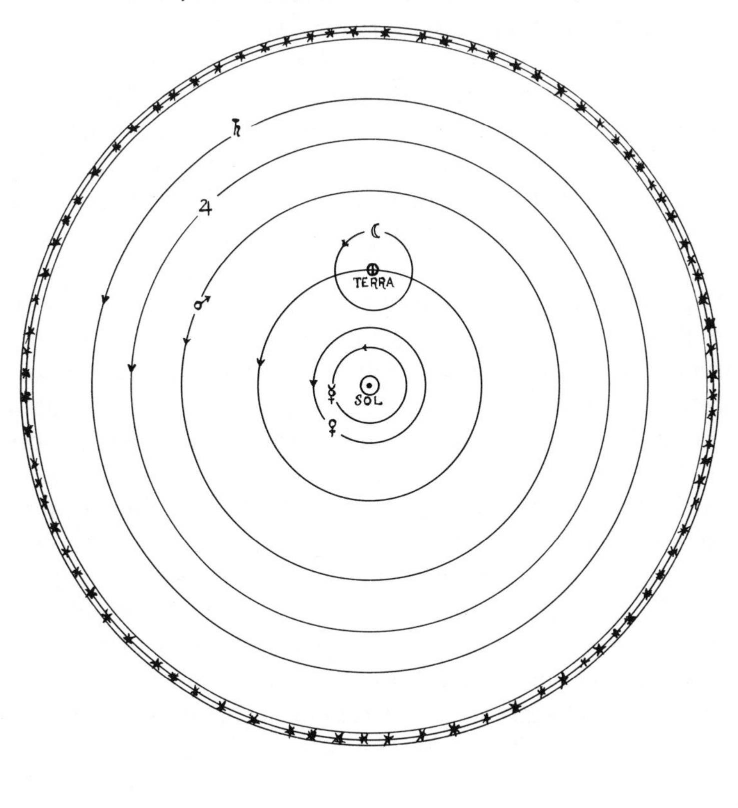

nor did it come up with more accurate positions. We now know the Copernican system was flawed: the planets do not move in orbits that are exactly circular, nor at constant speed. Copernicus himself was aware of this and attempted to correct the calculations by introducing small epicycles as Ptolemy had done. The flaw was removed later by Kepler, Newton and Einstein. But the seed had been carefully sown, Copernicus's sun-centered viewpoint was correct, and he gained immortality for his part in mindstep 3.

Ptolemy had pulled back the canopy to reveal divine points of light moving in space between us and the crystalline sphere of the immutable stars. In mindstep 2, earthlings were the passive audience and not a part of the show. Copernicus destroyed forever the secure feeling of a fixed, comfortably central, earth. It spun, it orbited, it moved with the sun and other planets through vast, unfathomable space. Compared to the greater cosmos, the earth showed as it truly was—a planet, one of many, an insignificant thing. Bright stars were huge, distant suns, and fainter stars were suns in their thousands, farther away. Nothing lay between us and mysterious infinity. The notion of the crystalline sphere was shattered. "So great, without any question, is the divine handiwork of the Almighty Creator!" exclaimed Copernicus in his 900 copies. Those were words ordered deleted by the Vatican censors. It was a disturbing revelation. It would take hundreds of years for greater humanity to adjust to the mindstep.

This was a mindstep that could be proved. But Copernicus did not prove it. He published only 27 observations instead of the thousands that were needed. Tycho Brahe was born in 1546 with a character well suited to the self-set task before him—exhaustive, detailed planet-tracking. Well educated in the Renaissance style, he had boundless energy and worked day and night on his labors. He forced assistants to work along with him at the same pace until they fell exhausted. He had bad traits, but even these were advantageous. Bombastic and quarrelsome, he argued with a fellow student at the university as to which one knew more mathematics. They fought a duel, Tycho's nose was sliced off, and so he wore a false one of gold and silver for the rest of his life. He had plaques and inscriptions put up in his house proclaiming himself the world's greatest astronomer. But his arrogance impressed the kings and princes who were to fund his research.

A strange series of events brought Tycho into astronomy. When he was 13 years old he saw a partial eclipse of the sun and was impressed. Then, on November 11, 1572, he was startled to see a new star in the northern sky. It was a supernova, an exploding star, a rare event in our galaxy. The remnants of that 1572 explosion can still be photographed as an expanding cloud of gas. Other people of the time saw the "new" star and speculated on its nature. Surely it was something close to the earth, perhaps closer than the moon, because it was so bright. Tycho decided to measure the distance. If it were closer than the moon the star would show a displacement (parallax) when looked at from different towns on earth. Tycho measured off the position of the nova from nearby fixed stars. There was no parallax; it had to be a very distant star.

Tycho wrote an account of his work, and his discovery caused consternation. Copernicus said the fixed earth was no longer fixed, and now the immutable stars were no longer immutable. Tycho lectured King Frederick of Denmark about the new star and its meaning, hinting that it was the duty of a civilized nation (such as Denmark) to support astronomical research. Then Tycho returned to Germany where he had spent his student days and started a rumor about applying for German citizenship. King Frederick summoned Tycho to return and asked him what he would need to do astronomy in Denmark. Tycho asked for the island of Hven off the coast at Elsinore, and sufficient money to build and operate a fine observatory. He employed the best instrument makers to engrave quadrants and circles for the ambitious planetary measurement program. Observations were made from balconies and turrets on the roof, and also from a basement room with instruments poking out through windows. He employed a large staff. The observatory was set in a landscaped park with 4 entrances, north, south, east and west. It was a palace, it was linked cardinally to the cosmos, and lord Tycho named it Uranienborg—Castle of the Sky.

Twenty years went by and Tycho and his assistants measured the positions of planets and stars in the heavens, collecting the raw data to test the theory of the moving earth. Twenty years of accurate, uninterrupted observations—never before available to science and all made possible by Tycho's drive and by Denmark's state support. But Tycho's forceful character had

Johannes Kepler (from a woodcut).

its negative aspects. His arrogance and overbearing attitude exasperated the Danish court, and on the pretext that Tycho had ill-treated one of the tenants on the island (no doubt a true charge), his financing was cut off. Tycho in a rage packed up his book of observational data and took his retinue to Germany. Uranienborg was looted and plundered, fell into decay, and there is now nothing left of the Castle of the Sky except the north-south line of the foundations.

As he wandered around Europe, Tycho carried a secret. His false nose was the clue. That student at the university *had* known more mathematics; Tycho was incapable of analyzing the data. He knew of the Ptolemaic and Copernican systems of course, and the dogma-based dispute about the moving earth. He had proposed a competing system, the "Tychonic," in which the planets go around the sun and the whole assemblage then revolves around the earth—a neat compromise. But Tycho needed a mathematician, otherwise his 20 years of labor would be wasted.

The most suitable person for the challenging task was Johannes Kepler, a brilliant mathematics teacher from eastern Germany who was already known for his skill in calculating the position of the planets. But old age was overtaking Tycho and his funds were gone. Kepler was underpaid and had no money to travel. How could the two men meet?

Kepler was born a Protestant and religion was a force within him. He was sensitive and emotional, a mystic and a sage. As a Protestant his salvation lay in the study of the Bible rather than in papal dogma. For Kepler the heavens were also a book written by God, to be studied as reverently as the Bible. In his mathematics he would work on a problem for years waiting for the day when divine inspiration would give him the solution. At the moment of discovery, of revelation, he would reach a high and weep for joy. It was Kepler's religion that was to bring him to the book of Tycho's data.

Rudolph II, emperor of Germany, invited Tycho to his court in Prague. Kepler was 200 miles away in Graz. The emperor offered Tycho the position of imperial mathematician, and Tycho, even though subdued by his ejection from Denmark and knowing that mathematics was beyond him, was still able to negotiate for himself a large salary. He took over Benatky Castle, set in the beautiful countryside near Prague, and immediately began to create a second Uranienborg.

Meanwhile in Graz Kepler was falling on hard times. With the swings of politics and religion Kepler was given the choice of renouncing Protestant theology and becoming a Catholic, or being banished from Graz. He could not renounce his beliefs. He had word of Tycho's new empire, and he craved to see the data. It seemed that the finger of Providence was pointing only one way. He packed his worldly possessions into two wagons and set off with his wife and family for Benatky. Tycho was overjoyed at the unexpected turn of events and gave Kepler and his family a warm welcome. Their meeting was charged with expectancy—one man had a planetary treasure chest, the other had the key.

There was continual uproar in the castle with carpenters, metalworkers and stonemasons rushing to the orders of their master. All the servants and assistants ate together Latin style at a large table with Tycho at the head. This paradigm was used to make the assistants feel inferior as Tycho poked fun at them, and to impress upon them the prestige and grandeur of their master. The data book was held by Tycho. On rare occasions it was produced and opened to a certain page for Kepler to see, but if he tried to turn the page and read more than his allowance, Tycho scowled. Before showing the entire book, Tycho wanted a guarantee that Kepler would prove the Tychonic system correct. But this, of course, was impossible. A theory proves itself. If it's right, it's right; if it's wrong, it's wrong. Kepler refused. He felt that destiny had brought him to the book and that destiny would open it for him. Tycho died, and Kepler took his place as imperial mathematician and began the analysis.

After 5 years of effort Kepler discovered the 1st law of planetary motion: the orbit of a planet is an ellipse with the sun at one focus. Then he discovered the 2nd law: a line drawn from a planet to the sun sweeps out equal areas in equal time intervals.

Copernicus was proved correct, the flaws in the theory put right. The truth was a strange break with the long accumulation of ancient beliefs. Instead of moving in perfect circles the planets moved in flattened circles—ellipses. This required a whole new geometry. The degree of flattening is measured by "eccentricity," e, and when $e = 1$ the orbit has maximum flattening. For the perfect circle of the ancients, eccentricity is 0.

Kepler found the sun was at the focus of the ellipse, a critical point in the geometry, which emphasized more than Copernicus had done the true importance of the sun. Ptolemy had been close to the idea of the foci of the ellipse with his "equants," but not close enough.

The 2nd law was a shock. Throughout the Ages people had believed in a constant steady speed for celestial bodies, but now Kepler revealed a peculiar pattern: a planet moved faster when it was near the sun. That was the gist of the 2nd law. How could this be? Kepler took it as evidence for a controlling influence from the sun, some force which was causing the effect. He was very close to the theory of gravitation.

One more law lay hidden in the figures of Tycho, and Kepler needed to work for 14 years to dig it out: the square of a planet's period is proportional to the cube of its average distance from the sun. Kepler's ambition in life had been to find a relationship between the distances of the planets, for then the harmony of God's work in the cosmos would be clear to him. It is therefore known as the harmonic law. Kepler went on to relate the numbers to a musical scale, as Pythagoras had done, and became more and more involved in the workings of the mystical cabala.

Because of the clash of religious forces over fine shades of theological dogma, Kepler was refused Holy Communion by both the Protestant and the Catholic churches. Ecclesiastically he was threatened with damnation, but he hoped to be reprieved in the afterlife on account of his earthly study of the solar system. He thought of himself as a priest of nature. He was buried in a modest grave in a graveyard in Regensburg. Two or three years later there was no sign of his grave for Swedish, German and Bavarian troops had turned the area into a battleground. When in 1773 the manuscripts, letters and diary of Kepler were discovered, they were purchased by Catherine the Great of Russia for the state observatory at Pulkova, and it is from a study of these documents that Kepler's full part in the mindstep has been established.

From Poland, Denmark, Prague and then to Italy, the step was endorsed in Rome by the most controversial figure of the scientific Renaissance—Galileo Galilei. Son of an impoverished noble, Galileo through his capabilities ultimately gained a life-tenured chair of mathematics at the University of Padua and was appointed personal philosopher to the Grand Duke of Tuscany. At rock bottom, though, he was practically minded, inventing the clock pendulum, the opera-glass telescope and the air thermometer.

Whenever Galileo gave a public lecture people flocked to listen. Whether he talked about the stars or comets, or the motion of arrows and cannon balls, he was always forceful and dramatic. The legend that he dropped weights from the top of the Leaning Tower of Pisa is probably true. He was born in Pisa, and it is said he did it to show the crowd below that a light object falls or accelerates equally as fast as a heavy one. (Only when one gets to something like a feather does the effect of air resistance show up to spoil the pull of gravity.) Galileo did this to prove wrong the ancient saying of Aristotle that a heavy object falls faster than a light one. Experimentation was his strong point. Book-learning was replaced with tests of nature and analyses—"ordeals" as Galileo called them. As events unfolded, this approach to the study of the greater cosmos was to generate an ordeal for him.

Galileo was a mixture of science experimenter and science writer, evangelist and plagiarist. He wrote and circulated a booklet entitled *The Messenger of the Stars* describing the latest dis-

15th-century Padua, Italy.

Galileo Galilei (from a woodcut).

coveries and theories of science. At one time the message of the Messenger was taken to be purely Galileo's, but a probing into Roman archives has disclosed no fewer than 4 sets of student notes with sentences word-for-word the same as his and with the same phrases underlined. It seems that Galileo and the students copied them from the same source, a professor Valla at the Collegio Romano. But in those days copyright was not an issue, and if a phrase was self-evidently true, those words were regarded as the property of everyone and authorship became blurred. One could even sell them in a Starry Messenger.

Galileo made discovery after discovery with his telescope and described it all in the *Messenger* with cosmological and theological comments. People were fascinated to read about mountains on the moon. Hadn't the Greek scholars in their ancient wisdom pronounced the moon to be smooth? (The Gilgamesh-Enkidu epic lay forgotten and buried under the tell at Niffar, and the English monks' sighting of a possible meteorite impact on the moon was ignored.) They were amazed to learn how planet Venus showed as a crescent like the moon. Aristotle had said Venus was self-luminous, but Galileo said the phases proved it shone by reflected sunlight. They did not know what to say of his claim that Saturn was accompanied by two "companions" on either side—Galileo's telescope was not powerful enough to show the rings. The sun showed dark spots on its surface, and from the movement of the spots Galileo argued that the sun rotated. If the sun was spinning, surely the earth could spin as well. Galileo's little telescope showed 4 moons circling planet Jupiter. Here, he argued, was a miniature solar system like the one hypothesized by Copernicus, and so Copernicus must be right. He shot down the old argument that said the earth must be fixed, otherwise the moon in its orbit would soon be "left behind." He said the 4 moons of Jupiter keep up with the moving planet, and our moon could do the same. It was Galileo, his telescope and the *Messenger of the Stars* that finally forced the Vatican censors to put Copernicus on the index of forbidden books. Galileo himself was ordered to desist from astronomical-cum-theological speculations and to go back to his experiments. He probably remembered the fate of philosopher-astronomer Giordano Bruno, burned at the stake for heresy.

Galileo gave his word of honor to the ecclesiastic court and

Aristotle, Ptolemy and Copernicus discuss the cosmos.

for 16 years he remained quiet, until an acquaintance of his was elected Pope Urban VIII. From previous contacts Galileo was led to believe that Urban appreciated the telescopic discoveries, was not unfavorable to the Copernican theory, and that the times were changing. He brought out something he had been working on for the past few years, *A Dialogue on the Two World Sys-*

tems, and submitted it for approval in the lower levels of the Vatican censorship bureau. The officials of the Inquisition read the pages, did not see the trick, and passed the work for publication. What Galileo had done was write a sort of play, a discussion between two professors on the merits of Ptolemy over Copernicus, and also in the cast was a man in the street called, in Latin, Simpleton. The eloquence of the professors was apparent in every speech in the book, and in the final scene Simpleton was persuaded that Copernicus was right. Unfortunately Pope Urban when he read the book identified himself with Simpleton and was enraged. Further publication of the *Dialogue* was stopped, existing copies were confiscated, and Galileo, although by now 70 years old, was summoned to Rome and forced to recant the heliocentric theory.

Galileo, his works and his trial by the Inquisition are still controversial. He died under house arrest in 1642, broken in body, stricken in mind, blind, unable to sleep and ridden with dropsy and ague. Would this have been his fate if he had been less abrasive, more diplomatic, more hypothetical? Would the Inquisition have censored Galilean astronomy if it had not been laced with Galilean theology? An ecclesiastic court, like any other court, follows a tightly confining set of rules, and its decisions are limited by those rules and by precedents. Within its own rules the Inquisition was correct. The sun-centered theory had been declared contrary to the Scriptures as far back as 1616, and Galileo had been officially warned not to support same. In the court's eyes he was guilty. A court judges the defendant, but the world judges the court. The real subject matter was the nature of the universe, not the words of the scriptures; mathematics, not theology. The Church was going too far in putting the mindstep on trial. As it turned out, the trial did more to spread the "heresy" than to stop it.

If the Church overstepped her boundary, then Galileo overstepped his. He was throughout his career strident and persistent in his attack on dogma, pushing science in an area it could not go. That was his choice. He did have other options. He could have kept quiet. He could have trundled off with his belongings like refugee Kepler, or sailed away in pursuit of religious freedom as the Pilgrims did 13 years before his trial. The day of the summer solstice came, Venus and Saturn retrograded, and the court found him guilty of heresy, ordering him to "abjure, curse and detest" this heliocentric mindstep. The evidence was impounded and there was no appeal.

Two hundred years later Napoleon Bonaparte conquered the city states of Italy, including the Vatican. French scholars followed the sword and took the full record of the trial back to France. Napoleon, in the spirit of freedom and human rights of the French Revolution, felt that Galileo had been unfairly suppressed. The Latin transcripts were carefully examined and then returned to the Vatican library by an agreement that had the world's first freedom-of-information clause negotiated into it. Henceforth French scholars must have total access to the Galileo file with a view toward vindicating him. Through the French studies we now know what took place, word for word, on that fateful, hot, Roman summer solstice.

As a result of pressure from France, the trial has been officially reopened by Pope John Paul II, but what the outcome will be in the long years ahead is anybody's guess. There are many questions in the reopened proceedings. Who is the defendant—the deceased Galileo or the flourishing mindstep? If it is the man, then he already has pleaded guilty to spreading a heresy even though expressly forbidden to do so by the court. If it is the mindstep, then is it to be judged in the 1633 time frame, or will new evidence be admitted? We now have witnesses who have seen with their own eyes the earth turning and moving in space.

Actually, as the transcript shows, Galileo did not do very well in defending the heliocentric theory. He testified how sunspots do not move across the face of the sun in straight lines, but follow elliptical tracks like the tilted lines of latitude on the globe. Therefore, he argued, the earth must be going around the sun in an orbit, because sometimes we are above the sun's equator, and sometimes below. Unfortunately for Galileo it would look just the same if the sun were going around the earth! With the solar axis sloping always in the same direction, the face of the sun would tilt toward the earth at one season of the year and away at another. What was his next point? He argued how ocean tides prove the spin of the earth. Because of the orbit the continents on the night side of planet earth move faster than those on the day side—the effects of axial spin and orbital motion add together on the night side to give an increased velocity— so as a continent swings into

Giordano Bruno, burned at the stake, 1600 A.D., for astronomical speculation (from a woodcut).

the hemisphere of sunlight it slows down and the waters of the ocean pile up against it (like a rear-end collision). Galileo's argument was a nonstarter—there are, as the Inquisition well knew, two tides each day, not one. Furthermore, the rise and fall of the ocean is synchronized with the moon, not the sun. The prosecutors were quick to shoot down these defenses. Galileo made matters worse by denouncing Kepler, who correctly had explained the tides as being caused by the moon. Said Galileo: Kepler has "given his ear and assent to the moon's predominance over the waters to occult properties and such-like trifles!" Galileo's famous line, allegedly said under his breath: "But it moves, nevertheless!" is not in the transcript. If he said it, it would make no difference. He was right in the natural laws of the cosmos, but wrong in the eyes of the court.

A French priest, Dominique Dubarle, proposes to have Galileo acquitted on some obscure technicalities of canon law. This may work. The Second Ecumenical Council has provided an umbrella under which Galileo could be pardoned. This Council declared: "Research performed in a truly scientific manner can never be in contrast with faith because both profane and religious realities have their origin in the same God." Yet even this sweeping principle has a catch in it for Galileo: motive is an issue, and the court must decide whether or not Galileo acted in a "truly scientific manner." In the retrial it might be found that he had dogma-overthrowing objectives, which again would be a crime.

Heretic or not, Galileo has an assured place in the halls of fame. He first explained the concepts of force and acceleration. He was on the verge of discovering gravity. His telescopes were effective, the first to be used astronomically and the first to be used in measurements. He attached a grid of ruled lines to the side of the telescope, and by looking through the telescope with one eye and at the grid with the other he saw objects against a measuring background. He discovered the 4 large moons of Jupiter and tracked them with this device. He invented an analog computer for predicting the positions as they would be seen edge-on in their orbits as viewed through the telescope. These 4 moons are called the Galilean satellites in his honor.

On the night of December 28, 1612, at 3:45 a.m., he discovered the giant planet Neptune. He plotted its position relative to Jupiter and labeled it as a fixed star, *b*. Bad weather followed and prevented observation, but on January 27 and 28, 1613, he saw Neptune again, and this time he did not label it "fixa." Star *a* was "fixa," but for star *b* there were motions! The point of light had moved with respect to the stars. It was a new planet beyond Jupiter, beyond Saturn. The frontier of the solar system had been pushed back 2 billion miles!

Galileo is honored for discovering Saturn's appendages, Jupiter's moons, lunar mountains and the myriad of stars in the Milky Way with his Tuscan optic, but he gets no credit for Neptune. This secret lay for centuries in the quill-pen notebook preserved in the National Library of Florence. It is clear from the pages that Galileo never followed up on his discovery, and he left it as simply a star that seemed to move. Perhaps even he was not prepared to accept a planet beyond the edge of the known solar system.

As he testified about the moving earth to the ten disbeliev-

A page from Galileo's astronomical diary showing Neptune. It says: "Beyond fixed star **a**, another followed in the same line, i.e., **b**, which was also observed on the previous night; but one saw motions . . ."

ing cardinals of the Inquisition under threat of imprisonment and torture, did Galileo hope those Latin words written in his diary would stay safely outside the courtroom? "Beyond fixed star *a*, another followed on the same line, that is to say *b*, which also was observed on the previous night; but one saw motions. . . ." Never in the field of astronomy had any human seen so far and so much through so small a telescope.

Isaac Newton was born on December 25, 1642, the year Galileo died. Within 2 weeks of his birth there was a partial eclipse of the sun, a total eclipse of the moon, and a conjunction of Jupiter and Saturn. But as a baby and a youth, Newton was not much interested in things cosmic. He was very good at making kites, sundials and model windmills powered by mice, and like other boys at the Grantham school he carved his initials deeply into the top of his desk. Usually he finished near the bottom of the class when the grades came out, but when he beat a local bully in a fistfight he decided to beat him scholastically as well, and from then on he scored top grade. His father died when Newton was in his teens and his mother needed him to run the family farm, but by the age of 16 it was clear he was not a farmer. He read mathematical books behind the hedgerows and did mechanical experiments. In one severe storm he neglected to secure the gates and barn doors and take the cattle to shelter. He was seen jumping into the wind and away from the wind, using his body to measure the force. So Mrs. Newton sent him off to college.

Newton was for the first time in an atmosphere of learning when he got to Cambridge University. He made a habit of scanning through the textbooks before the lecture course began, and then, with his total grasp of the material, he helped the lecturer along. The 3 years after graduation were the most intense in

Woolsthorpe Manor, near Grantham, England, and the tree where the lengendary apple fell, causing Newton to think of gravity.

Newton's life. During this period he figured out the force of attraction between the bodies in the solar system. He showed how Kepler's 3 laws were a direct result of and evidence for universal law of gravitation. This required the use of differential calculus, the bane of college math courses today, but not taught at Newton's college. He invented it. He had to, otherwise the gravity-orbit connection could not have been worked out.

He studied optics, showing how white light can be split into its 7 rainbow colors. Before that experiment people did not realize that white light is a blend of all the rainbow colors mixed together. They thought that white was a pure color and it was distortion that gave the other colors. Newton's findings were right of course; the other view is based on paint pigments, not light itself. Newton said light was made of corpuscles of different colors—precursor to the modern photon theory. Why 7 colors? This has always caused difficulty because most people seem to agree there is no way to separate out indigo and violet in the rainbow and really there are only 6. It seems Isaac Newton was color-blind and when he split the white-light beam he called a servant into the room and asked him what was there. "Red, orange, yellow, green, blue, indigo and violet," he said.

With mathematics and geometry Newton explained the cause of the tides in the oceans (not a continent-ocean collision), the reason for the bulge at the earth's equator (centrifugal force), and the reason for the precession of the equinoxes (gyroscopic action).

At the start Newton had to calculate the force of gravity at the surface of the earth: an object standing there is pulled slightly by everything else in existence—distant mountain ranges, the earth's core, rocks beneath the surface. When these forces are all added together the resultant pull is called the objects's "weight." The calculation involves that other bane of college math, integral calculus. Newton had to invent that too. By integration Newton showed that the earth acted as though all the mass were concentrated at one small point at the center, called the center of gravity. From there he could link the forces at the surface of the earth with the forces between earth, moon and other celestial bodies. Orbital theory came of age.

Newton brought mindstep 3 to fruition. The solar system and the stars in the universe were controlled by one set of physi-

Sir Isaac Newton on a British pound note.

cal laws, not the least being his universal law of gravitation. We now talk of "Newtonian" physics, so great was his impact. Albert Einstein later supplied the space-warp correction to account for anomalies near excessively massive objects like black holes, and in the close vicinity of the sun where a fast-moving planet is affected and the straight-line travel of a ray of light is disturbed.

Newton was eccentric. A confirmed bachelor, he did not care to speak to the house staff and they passed his meals to him through a slot in the dining-room wall. In his darkest year, 1693, he accused his closest friends of plotting against him, suffered insomnia and reported long, imaginary conversations. You are "endeavouring to embroil me with women," he wrote to philosopher John Locke in September of that year. Was he insane? Was he under drugs? One theory claims he was affected by metallic mercury absorbed during some of his ill-controlled experiments. Recently some hairs taken from Newton's head before death were tested for mercury poisoning by the nondestructive neutron activation process. The results: positive. Perhaps he was mentally ill. Workers in the felt industry fell prey to this illness in those days before environmental controls when the rags were processed in liquid mercury. Lewis Carroll portrayed the symptoms in the character of the Mad Hatter in *Alice in Wonderland.*

Another theory claims he suffered a nervous breakdown on the death of his mother. Perhaps the secret lies in the cells of his huge brain, now preserved in a jar in a medical museum. By all accounts it is one of the largest human brains on record, second only in volume, perhaps, to that of Akhnaton as estimated from the cranial cavity of Egypt's heretic pharaoh.

Whatever the cause of the malady, Newton recovered. He was elected a member of Parliament and was appointed Master of the Mint. He held the position for 28 years, and all British coins made between 1699 and 1727 in England and the American colonies were pressed under his general supervision.

Newton was a scientific genius and a man of deep religious conviction. At the end of his famous book *Principia Mathematica* he wrote:

> This most elegant system of the planets and comets could not be produced but by and under the contrivance and dominion of an intelligent and powerful being. And if the fixed stars are the centers of such other systems, all these being framed by the like council will be subject to the dominion of One, especially seeing the light of the fixed stars is of the same nature with that of the sun, and the light of all systems passes mutually from one to another. He governs all things, not as the soul of the world, but as the Lord of the Universe.

TEN
Upside-Down Cosmos

After Galileo, hundreds of astronomers took to the telescope. The universe was out there to be plumbed and fathomed. Galileo made his first telescope in May 1609 from 2 ordinary spectacle lenses. He put a long-sighted lens up front and a short-sighted lens at the rear. This was the way the telescope had been invented the previous year in Holland. Some say it was optician James Metius of cheese-town Alkmaar. Others say an assistant of spectacle-maker Hans Lippershey was bored with his job, held up an accidental combination of lenses to his eye and saw the weathervane on a steeple come close. Galileo took the idea and went on to grind lenses of higher power, finally reaching a magnification of 33. An object 330 feet away would then look as though it were only 10 feet from the eye.

As a wartime-evacuated schoolboy in the 1940s, 20 miles west of Newton's Grantham, this was the only way I could get access to a telescope. For one king's shilling I bought the objective (front) lens, and for a 2nd shilling the eyepiece. I taped them in the ends of a cardboard mailing tube. Everything I saw was new and exciting to me, but of course Galileo had seen it all, as the first person ever to see it, 333 years before. I beat Galileo on magnification, but it was a hollow success. My biggest telescope had a magnification of 60 with the lenses set in the ends of a spare galvanized iron drainspouting tube, and the whole mounted on a post in a cow pasture. The tube sagged under its own weight, morning dew clouded the objective and the shilling eyepiece lens dropped off into the grass. A cow must have eaten it. I couldn't afford a replacement. I followed the cows for days but the lens never showed up.

Kepler improved on Galileo's telescope. His idea was to use two long-sighted lenses (convex), one more powerful than the other. It was easier for the lens grinder to make a convex lens, and the field of view in the telescope was bigger. Also, crosswires for measuring could be placed between the lenses inside the tube, and everything could be seen with one and the same eye—not like the squint method of Galileo. The only trouble was that Kepler's design inverted the image, whereas Galileo's kept things the right way up. Inversion was no problem to astronomers, though. All drawings—and later all photographs—were reproduced upside down.

Newton improved things one stage further. He replaced the objective lens of the refractor telescope with a polished concave mirror. This overcame the main problem with refractions—blue light was bent more by the lens than red light and color fringes were produced. But in Newton's reflector, all colors were focused equally and the image was clear. There were disadvantages, though. The astronomer had to look down the barrel of the telescope with his back to the sky, and the image was still upside down.

William Herschel made excellent telescopes. Church organist by day, amateur astronomer by night, he set up his 6-inch reflector on the sidewalk in front of his house in Bath, England. Some of his eyepieces were no bigger than a fly's eye. He claimed he could get magnifications up to 2,000, and he ground and polished the lenses by a secret process. He was the astronomer mentioned by John Keats:*

> Then felt I like some watcher of the skies
> When a new planet swims into his ken;
> Or like stout Cortez when with eagle eyes
> He star'd at the Pacific—and all his men
> Looked at each other with a wild surmise—
> Silent upon a peak in Darien.

* *Fifteen Poets* (Oxford: Clarendon Press, 1941), p. 357.

John Milton had previously put his contemporary, Galileo, into *Paradise Lost:**

> . . . like the moon, whose orb
> Through optic glass the Tuscan artist views
> At ev'ning from the top of Fesole,
> Or in Valdarno, to descry new lands,
> Rivers or mountains in her spotty globe.

It was March 1781, and Herschel was carrying out an ambitious project, a systematic inspection of every star and nebulous object in the sky. On the 13th of the month, when he was examining a region in the constellation of Taurus, he came upon the new planet. He knew it was not a star because it showed as a disc and not a point, and it moved from night to night against the background of the stars. Since threshold zero there had been 7 wanderers; now there were 8. A musician-astronomer had broken the harmony of the spheres.

Herschel, a Hanoverian Britisher, named the planet after King George III, but French and American astronomers voted down the idea. By October of that year Cornwallis had surrendered and the star of George III was dim. They chose the name "Uranus," king of the starry universe in Roman mythology and father of Saturn. Even so, King George appointed Herschel to be Royal Astronomer (there already was an Astronomer Royal, and politically he couldn't be let go), at a salary of £200 per annum, with £4,000 more set aside in the British budget for the construction of a giant 48-inch telescope.

Casting a 48-inch mirror was a difficult task. Herschel tried to forget his previous unsuccessful experiences in casting a smaller, 24-inch mirror in the basement. The molten bronze, two-thirds copper and one-third tin, burst the mold and ran, red-hot, on the flagstones of the floor. The flagstones cracked and exploded. Workmen ran for their lives. With the biggest in the world, the 48-inch mirror, the molding was successful, but 12 men were needed to swing the polishing tool, each man pulling on a separate handle. Herschel had by now moved to the town of Slough, near Windsor Castle. Carpenters built a huge wooden pyramid behind his house, the latticework being needed to carry the tele-

* Ibid., p. 133.

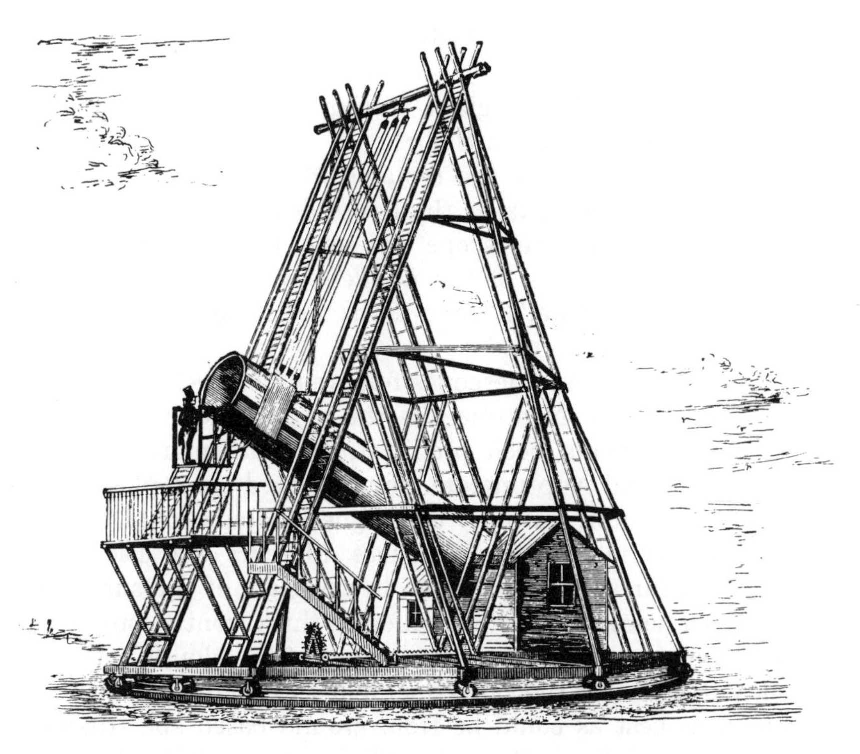

Herschel's 48-inch telescope at Slough, England.

scope and the observers' catwalks. The pyramid rotated on rollers on a brick trackway, powered by a man at a windlass. The tube of the telescope was 40 feet long and 5 feet in diameter. Visitors walked through it. Dukes, princesses and queens made the trip. George III led the faltering Archbishop of Canterbury by the hand, saying: "Come my lord Bishop, I will show you the way to heaven."

The telescope itself was a failure, though Herschel and the king would never admit it. England's climate allowed no more than 2 hours of good observing each week. Exposed to the night air, the mirror quickly tarnished and needed to be taken into the house and polished by the 12-man team every 6 months. Even under the best conditions Herschel did not see the details he had hoped for because of the atmosphere of the earth. The image, as with all high-powered telescopes, boiled and shimmered and the details were blurred. The optical dinosaur fell into disrepair and was finally dismantled by Herschel's son, John, who with his family sang songs in the giant rusting tube before it was laid aside in the back garden at Slough.

I was too late to see the remains when I went there in the 1960s. Wreckers were taking the house down to make way for a modern building. The telescope was gone, but I picked up a few square nails, mementos of the wooden pyramid, and a slab of marble from Herschel's front-room fireplace. The wreckers didn't object.

Bigger and better telescopes followed Herschel's in the 19th and 20th centuries. Neptune was the next planet to be discovered in 1846, having been predicted by mathematical calculations. Uranus and Neptune are cloud-covered giant planets, like Saturn and Jupiter, but only the vaguest details could be made out, owing to the immense distances. Mercury was also a mystery—small, far away and hidden in the glare of the sun. Venus showed as a pearly white disc, veiled in thick, continuous clouds. Astronomers measured the temperature of the clouds with thermopiles attached to the telescope, and the composition was determined as best as could be from ground-based spectroscopes. Some of the rainbow colors of sunlight were missing, being absorbed by gases in the atmosphere of Venus above the clouds. One absorption band showed carbon dioxide, another showed water vapor, but the latter measurements were affected by water vapor in the earth's atmosphere.

Earth's atmosphere was the astronomers' main obstacle. It absorbed, deflected and blurred the light which was then their only link with outer space. Desperate attempts were made to get clear. In the 1950s, U.S. Navy commander Malcolm Ross and Charles Moore took off one day from South Dakota in the gondola of a stratosphere balloon. Moore was a physicist and his task was to point an electronic spectrograph at Venus when the balloon reached sufficient altitude. At 80,000 feet the balloon was clear of 98 percent of the earth's air. The huge plastic bag swayed and lurched and twisted, and the temperature went down to −40°F. They went down too, and batteries and equipment were jettisoned to slow the fall and save the fluttering plastic above them. The gondola landed on a slope and rolled. The parachute got caught by the wind, and gondola and occupants bounced along for a good part of a mile. Fortunately they were unhurt, and the scientific records from the top of the atmosphere were saved.

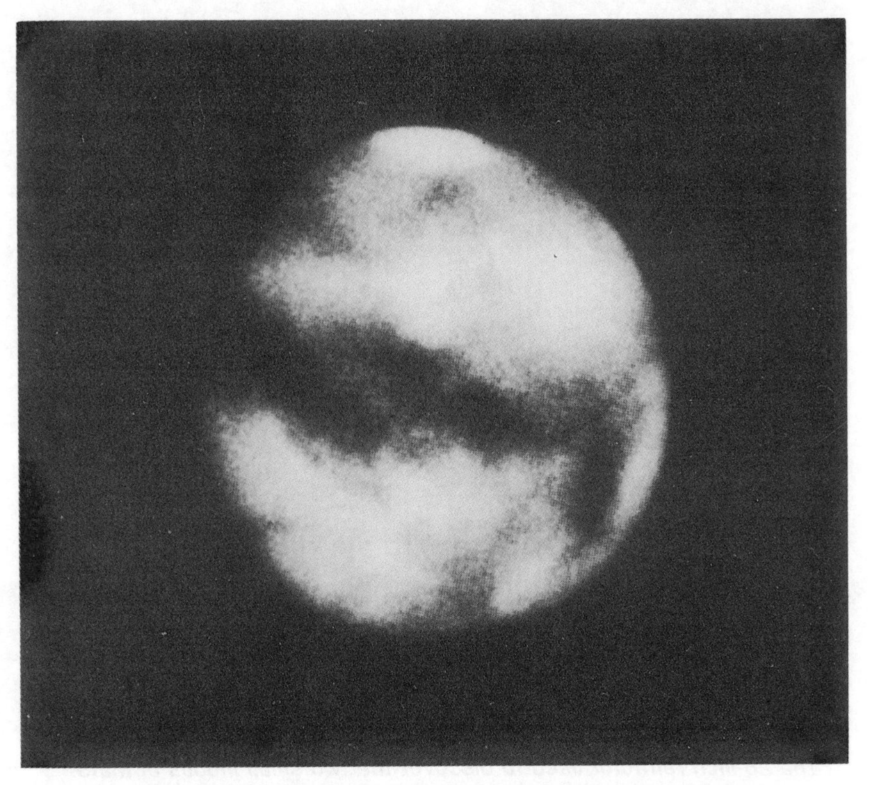

Mars, south at the top, showing Syrtis Major to the left.

The only planetary surface within reach of the telescope was the surface of Mars. Astronomers saw large expanses with an orange hue, and dark markings which rotated with the planet. The bright areas were the so-called red sands of Mars, but the dark areas were a mystery. Even the color was in dispute. Some said olive green, others said gray or blue. Part of the problem was the false color fringes caused by the telescopes, and part was the reddish glare of the desert regions which upset the estimates of color. There was agreement on the whiteness of the polar regions of Mars, which were thought to be snowfields like the Arctic and Antarctic regions of the earth. Occasionally a yellow cloud of dust stirred from the desert to move over the surface at 30 miles per hour as a dust storm. Some years astronomers found the entire planet engulfed in a storm which obscured all details of the surface.

The 26-inch refractor used to discover the two small moons of Mars.

The strangest happening in the study of Mars was the report of the "canals." Astronomer Giovanni Schiaparelli was carefully mapping the deserts and dark regions. The image of Mars is hardly ever steady, due to disturbances in earth's atmosphere—it's like looking at something at the bottom of a pool when swimmers are in it. Suddenly there was a brief instant when the atmosphere became still, and Schiaparelli saw lines crossing the desert. He hardly believed what he saw, but he caught more glimpses later on in the night. He announced his discovery in Italian, using the word "canali." That word has two meanings: natural channel or man-made canal. The news media ran the word "canal," and so the legend of the canals of Mars was born.

Bostonian Percival Lowell took up the question of the canals in a big way. He was not a professional astronomer but was well versed in mathematics and physics. His interests ranged from flowers to Shintoism, and with his own money (he was a man of private means) he built an observatory at Flagstaff, Arizona, dedicated to the surface of Mars. It was named after him, and his body lies buried on a hillside near the telescope dome. During his lifetime he mapped hundreds of canals that he said were bringing water from the poles to the oases in the desert. Some of the canals were double for a return circulation of the water. The Martians were irrigating the desert!

Other astronomers could not see the canals, and they said so. Lowell claimed their eyesight was not good enough, and their telescopes inferior. But under ideal conditions and high magnification Lowell's canals showed as a disorderly array of small dark patches. The canals were an optical illusion.

Beyond Mars there is a zone of asteroids. Instead of a single large planet, smaller bodies formed. Some of these collided to make more fragments, and these are deflected by Mars and fall to the earth's surface as meteorites. Jupiter was the cause of this abortive planet. Circling around with a strong gravitational field, Jupiter tore the swirling proto-planet apart as it was forming at the beginning of the solar system, approximately 5 billion years ago.

Jupiter is the largest planet in the solar system, with 1,320 times the volume of the earth. It is a pretty sight through the telescope with its colored cloud belts and its 4 famous moons. The spectroscope shows that certain wave lengths of light are absorbed, and that ammonia and methane must be in the atmosphere. By watching stars through the edge of the atmosphere, astronomers found that the bulk of the atmosphere was made up of the lighter molecule of hydrogen. Not a pleasant mixture—hydrogen and poisonous pungent gases. Storms on Jupiter can be seen with even a modest telescope. Circling the equator is a bright zone of clouds, separated from the south tropical zone by a dark belt. Then there is the "Great Red Spot." Robert Hooke saw it first in 1664, and it has been seen on and off ever since. "Spot" is an understatement. It is 30,000 miles across, room to hold 3 planet earths side by side. The spot may be even larger in the regions beneath the cloud-tops. Earth-based astronomers could never agree on what it was. Theories ranged from a permanent cyclonic storm to the glow of a giant volcano beneath the clouds. One theory said the object was a disc-shaped island, floating half-submerged in a dense, semiliquid atmosphere.

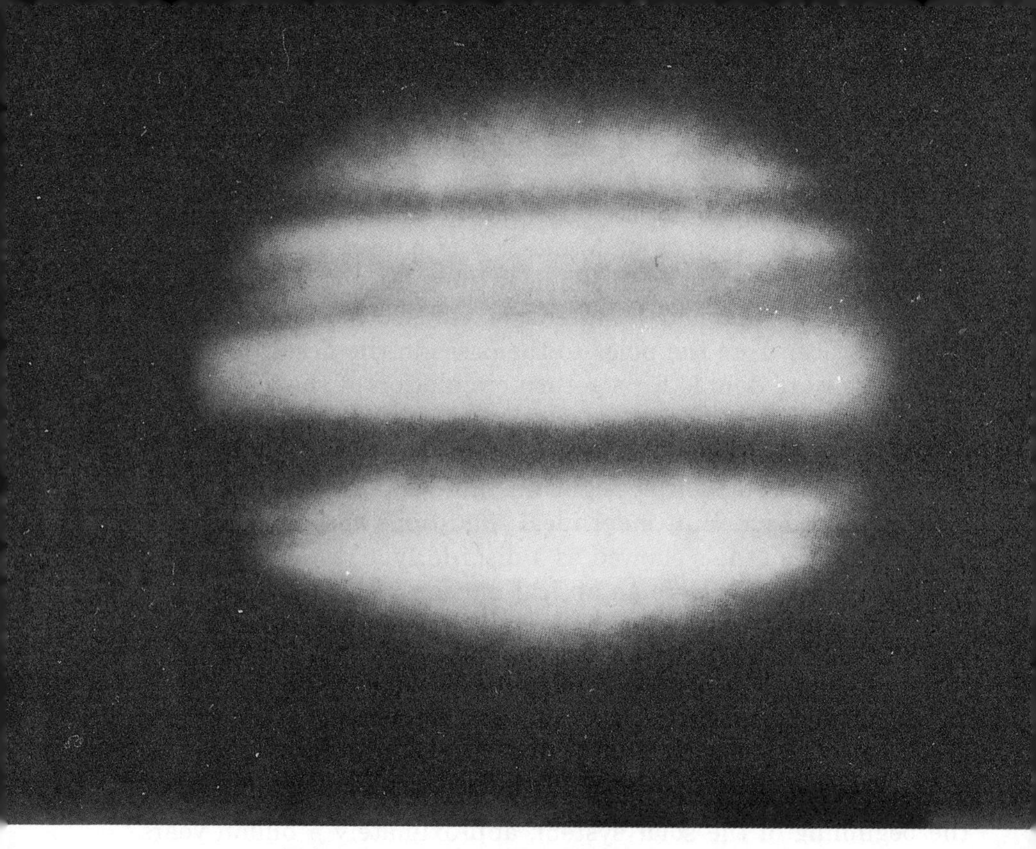

Jupiter through the 100-inch Mt. Wilson telescope.

Saturn through the 100-inch Mt. Wilson telescope.

If Galileo was pleased with what he saw with Jupiter, he was perplexed when he turned his telescope to Saturn. Instead of 1 planet he saw 3 set out in a row, and the large companions didn't move like the moons of Jupiter. Year by year the companions grew smaller until they vanished. Galileo could not believe his eyes. The honor of solving the problem of Saturn's "companions" goes to Christian Huygens of Holland, who was the first to recognize the rings. "Aaaaaaa, ccccc, d, eeeee, g, h, iiiiiii, llll, mm, nnnnnnnnn, oooo, pp, q, rr, s, ttttt, uuuuu!" he wrote in 1656. A coded announcement was not unusual in the 17th century; it allowed the astronomer time to verify a revolutionary finding. Three years later Huygens was sure of what he had seen, so he published the decoded anagram. The letters when rearranged spelled out a Latin sentence which declared: "Saturn is surrounded by a flat ring inclined to the ecliptic and nowhere touching the planet."

Earth-based observers see the rings from the side, and as Saturn moves around the sun every 29 years the rings seem to open and close. When the earth was exactly edgewise to the rings, Galileo could see nothing. When the rings were open, his imperfect telescope showed them as 2 side companions to the planet.

Giovanni Cassini first noticed the dark circle which separates the rings into 2 main portions. It is called Cassini's Division after him. Inside the Division, telescopes showed a bright ring, and outside the Division a fainter ring. A 3rd and fainter ring could just be made out with large telescopes. It was called the "crepe ring" because of its semitransparency.

Saturn is a giant planet, like Jupiter, with a heavy atmosphere, bright zones and cloud belts. Markings are not so well defined as on Jupiter, and there is no "Red Spot." Methane and ammonia bands showed in the spectroscope, but the lines were weak because of the low temperatures out at Saturn's distance. It was theorized that most of the ammonia had been frozen out of the atmosphere to make the crystal particles of the clouds and the heavy sludge that must lie below them. The low temperatures may also be the cause of the faintness in the colors of the cloud belts. Like Jupiter, Saturn has concentric shells of ice surrounding a central core of rock and iron.

Saturn was the end of the solar system cosmos until the discovery of Uranus and Neptune. Then, beyond Neptune there was found a 9th planet, Pluto the iceball, not much bigger than earth's moon. It was discovered by a mathematical fluke. Uranus and Neptune were off course, lagging slightly in their orbits. A 9th planet was suspected as the cause of the disturbances. By the early 1900s, astronomers had gone so far as to pinpoint the constellation where the planet was. The sky was thoroughly photographed, hundreds of images examined, and in 1930 Clyde Tombaugh found a 15th-magnitude "star" that moved from night to night: Pluto.

So much for the mathematics. Orbital theory had blazed the trail to the missing planet, and after a laborious search it had been found. Tombaugh, just turned 24 and employed at Lowell's observatory only on a trial basis, was immediately promoted to the permanent staff. He later went on to discover 775 asteroids, one of which is named after him. But Pluto could not have

Pluto through the 200-inch Mt. Palomar telescope.

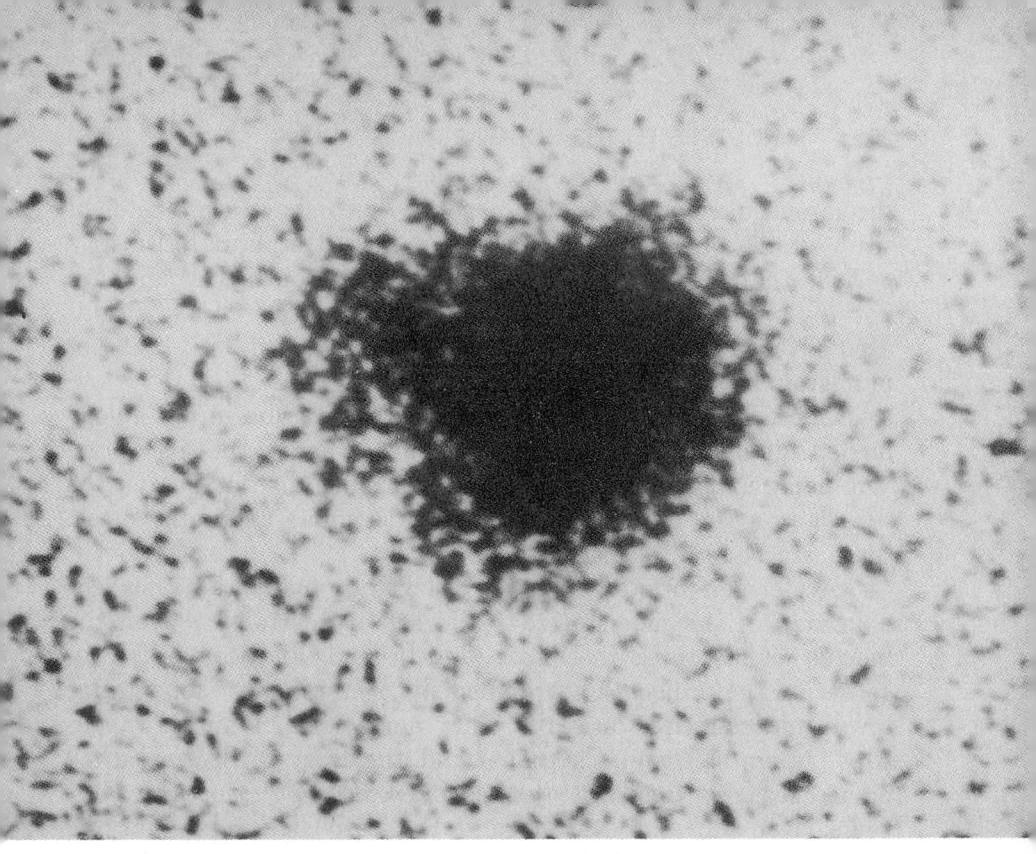

Enlarged photographic image of Pluto showing its moon, Charon.

caused the lag in Uranus and Neptune. Exact measurements show that it is too small. Its diameter is only 1,500 miles—smaller than Mercury, one-tenth the mass of earth's moon. To cause those lags, Pluto should have been a thousand times bigger. Conclusion: there must be another planet out there, planet X.

Ultimately Pluto may not qualify as a true planet. Its orbit is not circular and it comes inside the orbit of Neptune. From now until 1999, Neptune is the farthest planet from the sun, not Pluto. This smallest of the planets seems to be made of ices, more like the composition of the nucleus of a comet. Its surface is solid and cloud-free, probably made of frozen methane. Perhaps Pluto was originally a moon of Neptune, but was pulled into a solar orbit by a close encounter with planet X.

On the other hand Pluto, though it is small, has an even smaller moon. This brings it back into the realm of true planets because there is no other case in the solar system of a moon with a moon. Charon, Pluto's moon, was discovered in Arizona during a month of exceptional clarity of the atmosphere in 1978 when the minuscule photographic image of Pluto showed a cluster of silver grains on one side. The bump moved around Pluto in a period of 6 days, the length of the Charonic month. We can learn nothing more about Charon from earth-based telescopes. A hundred silver grains on a photographic film. The symbolic limit for mindstep 3.*

Charon is named after the bearded sailor who in mythology ferried the dead souls on their last journey across the river Styx. The name is appropriate. At these enormous distances from the sun where gases are frozen and the earth shows as a faint speck, there can be no life in any form.

A German brewer, Johannes Hevelius, made the first reliable map of the moon in 1647. He set up his telescope on the balcony of his house in Danzig, and he and his wife worked together on the project. They named the craters on the moon after towns and countries on earth. They named the mountain ranges Alps, Apennines and Caucasus, and the large dark areas they named seas. To them they looked like cold, watery oceans. Four years later Giovanni Riccioli and Francesco Grimaldi, working from Bologna, named the craters after persons. Tycho, Kepler and Copernicus were there, and many influential people of the late 17th century were flattered to find themselves on the moon map. Two of the smaller features were named after Hevelius and Galileo. Two of the larger ones were named Riccioli and Grimaldi. The names have stuck, and craters, even on the far side, are named after persons. Hevelius, however, won out with the mountains and mare (seas).

As with the planets, the moon maps were published upside down, with the south pole at the top. The familiar shape of the face of the man in the moon is made up of the Sea of Rain and Sea of Serenity for the eyes, and Mare Nubium for the mouth. The narrow Sea of Cold, Mare Frigoris, gives the furrowed brow. But Hevelius would have done better to have called them plains, not seas.

Meteorites fall to the earth, and they must also be hitting the moon. Every thousand years or so the earth is struck by an

* This silver-grain symbol of mindstep 3 actually was produced in the mindstep that followed. The momentum of an old mindstep tends to continue into the new—artists even now make mindstep 1 representations of the sun and moon; Ptolemy's system of mindstep 2 finds a place in the computer programs in the Appendix.

object weighing 40,000 tons or more. Quite a blow. Earth's atmosphere protects planet earth, but on the airless moon a crater 1 or 2 miles in diameter is produced. Thousands of craters have been blasted into the moon's disc over the long ages of the past.

The crater Copernicus is a showpiece. The telescope's-eye view looks downward into a vast amphitheater. Shadows reveal craggy walls 11,000 feet high, and the bowl is 56 miles across. Landslides developed after the crater was made, and a series of terraces can be seen on the inner ramparts. Tycho, a magnificent crater and probably the one most recently formed, is near the south pole. It is the jewel at the throat of the profile of the lady in the moon. Like Copernicus, it was blasted into place by the impact of a large meteorite. A body 2,000 feet in diameter with a mass of 450,000 tons would have been sufficient. From the present rate of fall of meteorites on earth the moon would be hit that way once every 10 million years or so, but the bombardment may have been more severe in the past. Tycho must be a young crater because of the remarkable system of rays. These were formed by the dust and debris spewed out from the original impact. The rays are bright because the dust contains solidified droplets—rock glass—which reflects the sunlight like cat's-eye reflectors. The dust has settled on craters, mountains and seas, stretching 2,000 miles down (or up?) to the Sea of Serenity. Tycho must be younger than any feature so marked by the rays.

Before the Sea of Rains was formed the surface in that part of the moon was rough and mountainous with a few scattered craters. A meteorite about 10 miles in diameter crashed into the surface a little north of the present center of the sea. For a micro-instant there was a clean, punctured hole, then all the energy was unleashed. The solid crust was torn away to expose a crater. Debris shot out in all directions in a blinding flash. Some pieces may have reached the earth as meteorites. Gigantic blocks were hurled across the lunar landscape from the seat of the explosion, plowing through mountain ranges and craters on the way. The scars are still visible, even though this all took place at the beginning of lunar time. Lava seeped up from the interior to cover the floor of the crater, and smaller meteorites cratered the crater floor.

Ptolemy in his ordered cosmos paid no attention to comets. They did not fit the epicycles. Kepler believed they moved in straight lines, unpredictable and uncontrollable. Comets showed up unexpectedly in the night sky, moved silently as harbingers of disaster, then disappeared into the deep. They were demons, monsters, evil spirits, occult signs. Many of the legends of swords hanging in the air can be explained as the curved dust tail of a comet shining in the far sunlight. Comets carried the aura of mystery of mindstep 1 into mindsteps 2 and 3.

Plutarch wrote about a brilliant comet which fittingly appeared when Julius Caesar was assassinated. Aristotle said a comet caused the earthquake at the Greek cities of Helice and Bura. The Aztecs feared Cortes because of the comets that preceded his landing at Veracruz. William the Conqueror used Halley's comet, one of the largest known, in psychological warfare. The comet appeared just before the Battle of Hastings in October 1066. King Harold's opposing army said it was a bad omen and the sky gods were against them. William spread the word that this comet was different. It was a sign from heaven of a conquering sword, assurance of glorious victory. William won the battle, of course.

In the 19th century, French wine makers believed that comets were hot objects and the heat produced good grapes. Years when a comet showed up were therefore good vintage years. Catalogs advertised "superb comet wine."

To the unaided eye a comet seems to have a hairy head and tail. Through a telescope it dissolves away as faint, wispy gases, and sometimes a pinpoint nucleus can be seen. The head of gas, glowing in the sunlight, is sometimes a million miles in diameter. Gases streaming away from the head make the comet's tail, which can be many millions of miles in length. There is often a straight tail made of glowing gas, and another curved tail made of dust and debris.

The size of the nucleus can be figured out from the amount of sunlight reflected from its surface. It is usually less than a mile across. Even the greatest comets seldom have a nucleus more than 10 miles in diameter. The nucleus is a mass of ice, dust and rocks. Ten miles is small in terms of the solar system but large from the point of view of impact with planet earth. As the nucleus moves in toward perihelion near the sun, the heat vaporizes the ices and jets of gas are blown out by the force of the brilliant sunlight. These gases ionize and glow, and the tail points away from the sun. As it crosses the earth's orbit the ball of ice and rock is transformed into a majestic object. The tail grows until it is an enormous plume, sometimes stretching halfway across the sky. The head puffs up, new jets of gas spurt out, and the comet changes shape from night to night. As it changes it seems to flutter in slow motion in the solar wind.

Comets are named after the several discoverers' and the International Astronomical Union allows a maximum of 3 names per comet. Kaoru Ikeya, a lathe operator in a Japanese piano factory, determined at age 17 to bring honor to his family name. He ground his own telescope mirror and hunted for a comet. After a 2-year search he found one and it carries the name Ikeya as he planned. His hometown bathed in reflected glory and set about converting an abandoned water tower into an observatory for him. From there he went on to discover 2 more comets.

Halley's comet breaks the rules. It has come to the sun regularly throughout recorded history, but nobody knows the name of the discoverer. Using the new gravitational theory of his friend Isaac Newton, Edmund Halley computed the orbits of the bright comets of 1682, 1607 and 1531. The orbits were the same—an enormous ellipse going out far beyond Saturn—and he realized those 3 apparitions were caused by the same object. Comets were not sky-monsters; they were a part of the order of the solar system, benign and predictable. If I am right, Halley said, "it should return . . . about the year 1758." It did. The comet was picked up in the telescopes on Christmas Day, 1758, and was a spectacular sight through the spring of 1759. Halley did not live to see his comet return. He died in 1742, but a romantic French painting shows an angel raising Halley from the grave to look at "his" comet.

Halley's comet is one of the largest, with a nucleus about 10 miles across and a tail that can stretch to 50 million miles. In 1910 the earth passed right through this tail. Mindstep 1 calamities were predicted—fire, suffocation, earthquakes—but nothing bad happened. The gas was so rarified it did not even affect the earth's upper atmosphere. This most famous comet has its next rendezvous with the sun in 1986, and the countdown is on schedule. Then, after this 12th encounter with the sun since William the Conqueror's manipulation, Halley's comet swings out again in its curved track across the orbits of earth, Mars and Jupiter, tail-first to the blackness beyond Saturn.

On June 30, 1908, in the desolate swamps of Siberia, people within a radius of 400 miles saw and heard a tremendous explo-

Halley's comet at the 1910 return.

sion. A brilliant object moved overhead from the southeast, dropped sparks and left behind a trail of smoke. The sound of the explosion reverberated 700 miles away, and many animals were knocked over by the blast of air. A man at the Vanovara Trading Post experienced the effects of a searing heat, was lifted from his seat and thrown several yards. Earth tremors followed, houses shook, windowpanes broke and household things began to move and fall. An engineer on the Trans-Siberian Railroad stopped his train because he feared it was going to be thrown off the tracks by the heaving, shaking earth. A catastrophic event had taken place far to the north, but people couldn't penetrate into the swamps to find out what had happened.

It was 19 years before the first scientific expedition broke through. Over a 30-mile area in the Tunguska River Valley trees had been knocked over and charred by a fireball blast. Reindeer and all other wildlife living in the area had perished during that brief instant 19 years before. A comet nucleus had struck the earth.

Fortunately events like this are rare. Halley's comet is on a track well clear of the earth. None of the periodic comets are currently a threat. But comet collisions have done tremendous damage in the past. The Biblical towns of Sodom and Gomorrah may have been destroyed by comet impact. Then too some 65 million years ago there was a geological event which covered the earth with a layer of iridium-rich dust. This coincided with the end of the Cretaceous period and the beginning of the Tertiary. Whole species disappeared, including all of the various types of dinosaurs. Was it a comet impact? Were those large, seemingly dominant reptiles killed off by something from the cosmos?

Iridium is a silvery gray metal, rare on the earth's surface, but more concentrated in the iron of meteorites.One theory for the extinctions is that a 6-mile-wide, metal-bearing meteorite, or comet nucleus, smashed into the earth. A computer simulation shows how a plume of thick dust spread out in all directions from ground-zero impact, blotting out sunlight for weeks, perhaps years. All the large, leaf-eating, cold-blooded reptiles suc-

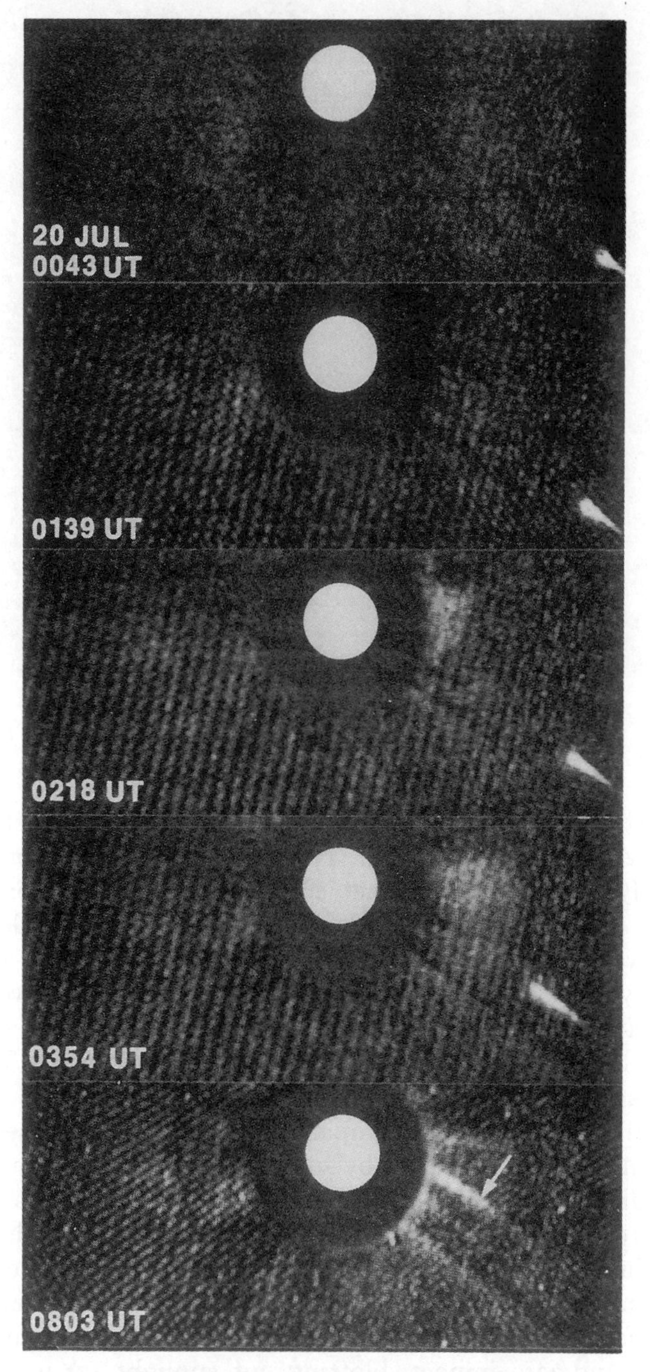

A comet collides with the sun, 1981.

cumbed. The 150-million-year reign of the dinosaurs ended. Smaller animals, like the insect-eating shrew, survived, and during the next 65 million years the mammals took over the earth. There are other theories, but this one explains most of what seems to have happened at that time. These catastrophic collisions occur on the average once every 100 million years or so. The Tunguska event was, fortunately, not quite so catastrophic. When a collision takes place on the airless moon, the crater stays there for all time. But when the impact takes place on earth, the geological processes of erosion slowly heal over the scar.

In the desert of Arizona there is a raw gash nearly a mile in diameter. An enormous mass of iron hurtled through the air some 5,000 or 10,000 years ago. The explosion forced up the rim of the crater into an escarpment 160 feet high; huge boulders were thrown out of the crater to lie in a jumbled mass around the rim. Intense heat turned the sandstone into a glassy foam, and pressure pulverized the rock beneath the crater floor into a fine powder. It will be millions of years before erosion can remove this scar.

The fragments of iron from the Arizona meteor crater contain diamonds, about 10 small diamonds per pound of meteorite. This shows how the meteorite solidified under very high pressure. The large iron crystals show there was a long period of cooling in the parent body.

Fragments of stone meteorites contain the same chemicals and some of the same minerals that are found on earth, but they are a very unusual mixture. Somehow in the past the various crystals of feldspar, olivine, and other minerals have been splintered and crushed and then packed together again. Through a microscope a stony meteorite looks like a piece of terrazzo cement, or a modernistic tile pattern. Embedded in the mass there are spherical mini-pebbles called chondrules, and tiny flecks and veins of pure metal. This material has probably not changed since the time when the solar system began.

Meteorites fell to the earth's surface throughout each of the mindsteps, and earthlings slowly adapted to these events. By the time of mindstep 3, iron and stone meteorites were recognized for what they were—tangible samples of the cosmos, things from outer space.

In Mindstep 0, earth and sky were one. The goddess Ge was

everything. People in those times probably believed that stones could grow out of the ground like plants growing from a seed, and it was no more surprising to have stones materialize from thin air. The Omphalos was a sacred stone held together by bands at Delphi in ancient Greece. This stone was said to be the center of the flat, disc-shaped earth. Delphi was the place where the famous oracle-maiden answered questions about the future. The temple was originally dedicated to the earth-goddess, Ge, and the sacred stone which gave the oracle her power was a meteorite. Later, as mindstep 1 developed, the temple was dedicated to the sun-god, Apollo.

On the larger scale, catastrophic explosions were all manifestations of "mother earth," whether they were volcanic caldera or astroblemes. There was no way for ancient people to know the difference. American Indian legends vaguely refer to the Arizona meteor crater as a place to be feared and avoided. The dry riverbed nearby is called "Devil Canyon."

When earth and heaven became truly separated in mindstep 1, any object falling from the sky was clearly a spirit or god-like thing. The tribes of the Arabian Desert before the time of the holy Prophet built the black stone into the structure of the shrine at Mecca. In Mesoamerica archaeologists unearthed a meteorite buried as a spirit and wrapped in mummy cloth. The Biblical Ephesians worshiped a "thing that fell from the sky."

But in mindstep 2 there was no logical place for meteorites in the order of Ptolemy's system, even though these objects were an important clue to the physical nature of those mysterious lights orbiting between earth and the starry sphere. Any reports of same were dismissed as ignorant superstition. The Omphalos and the piece of iron that King Arthur pulled from the stone were pure legends, inventions of the mind. But the objects continued to fall. The pressure of the hard facts built up.

By the time mindstep 3 was halfway through its course there was no denying the true nature of these extra-terrestrial objects. A great debate took place in the 18th century with many pros and cons. It ended abruptly in 1803 when a shower of hundreds of stones fell at L'Aigle, near Paris. A large meteorite entering the atmosphere had fragmented under the heat and pressure of the encounter. A special committee of the Paris Academy went out to investigate. Conclusion: it was all extra-terrestrial materi-

A transit telescope, used for time keeping.

al. Chunks of debris were going around the sun as the earth was. A few years earlier the French satirist Voltaire had resisted the new idea. He is supposed to have said: "I could more easily believe that an astronomer would lie than that stones could fall from heaven!" He died unconvinced, but mindstep 3 went on.

ELEVEN

Stars and Galaxies

The sun's galaxy is a collection of 100 billion stars suspended in the darkness of space. There is a bright bulge at the center, and around it are swirling arms of stars and clouds of dust and ejected gas. It takes a ray of light 100,000 years to pass from one edge of the disc to the other, and the photon speeds at 186,000 miles per second. This lumbering, cosmic pinwheel rotates once every 200 million years or so, with stars at the edge moving more slowly than those at the center. Beyond the galaxy of the sun there are others spreading into the distance to unfathomable limits.

Mindstep 3, the age of revolution, was also the age of the telescope. Each new optic showed more galaxies, more wonders, and the frontier was always beyond the astronomer's ken. At the outset Copernicus had already hinted at the immensity of the cosmos—"So great, without any question, is the divine handiwork"—and with each new discovery planet earth seemed smaller and less significant. Even though stellar astronomers broke with the upside-down planetary observers and printed photographs the right way up with north at the top, the cosmos was no easier to understand. It was a revolution against all previous thinking for earthlings to have to admit to playing such an insignificant role in the vastness of the cosmos.

Hernando Cortes burned Aztec nobles at the stake in the square of old Mexico City in synchronism with manifestations of

By the 1940s the astronomer was riding inside the telescope—the 200-inch, Mt. Palomar observatory.

the planet Venus. Eighty years and one mindstep later the Italian philosopher Giordano Bruno was burned at the stake in the Field of Fiori in Rome by the Inquisition. By chance, Venus was again brilliant in the morning star aspect. One of his crimes was astronomical heresy, and he was found guilty after a 7-year trial. He had written about the boundless size of the cosmos and the immortality of the human soul within it. He proposed that there might be an infinity of worlds. The authorities who condemned him were hung up, it seems, on a theological fine point: God could not create an infinite universe, therefore anyone who said an infinite universe existed denied the existence of God. Creation in the eyes of 15th-century theologians was a limited, graspable process—first the clay of the earth, then the fire of the sun, then the jeweled stars. There must be ponderable order, a place for humans and a function for God. To whisper infinity was to challenge the structure.

Galileo was mindful of Bruno's fate. Like Bruno, he lectured on philosophy and was known for his revolutionary memes. Ten years after the burning he saw through his new telescope what no one had seen before: an infinity of stars. He split the Milky Way into pinpoints, he saw double stars and clusters, and perhaps even the faint glimmer of the spiral galaxy in Andromeda, the twin of our own galaxy, but he played these discoveries down. Through the telescope he saw the marvels of the night sky; on the earth he saw the absolute power of the Inquisition.

Today we enjoy the full benefit of the mindstep 3 revolution. We can look at the wonders of the material cosmos and take it for what it is, regardless of heresy. In the evening of late summer in the northern hemisphere the Milky Way stretches as a pale arch through the sparkling constellations of Cygnus and Aquila down to Scorpio and Sagittarius where it divides into two "rivers." The inky blackness that divides it is an obscuration, a dust cloud hiding the very center of the galaxy. Through binoculars and high-powered telescopes the glow resolves into stars, but the dark regions seem like a void. The stars are the young stars of Population I, stars of the spiral arms, stars like the sun. The most powerful telescopes reveal a spherical halo above and below the disc of the arms, and this halo is made up of the older, fainter stars of Population II.

Only one-seventh of the young stars are single. The majority are found in close groups—double, triple and quadruple star systems—bound together by gravity. The sun is in the minority. It does not have a companion.

Alcor and Mizar are a good example of a double easily seen with the naked eye. Mizar is the star in the middle of the handle of the Big Dipper and it is separated from faint companion Alcor by one-fifth of a degree. The actual separation is 13,000 astronomical units (1 AU is 93 million miles, the average distance of the earth from the sun), and Alcor will take a million years to go around Mizar. It is one of the widest pairs known, and was there to be seen as soon as earthlings left the caves and looked up to

The Milky Way in Sagittarius.

the heavens. Mizar itself is a close double, resolvable only with the aid of a telescope. Italian Giovanni Riccioli discovered this when he turned his telescope from the moon to Mizar in 1650.

The star epsilon Lyrae is a double-double—the 2 stars seen with the naked eye are each split in the telescope. Each pair spins like a normal binary, and the pairs revolve around each other. Castor in Gemini is a sextuplet. Each of the 2 stars seen in a telescope is a whirling twin, and there is another fainter twin system moving in a large orbit around the foursome at the center. Multiplet systems divide into pairs. There is no case of 2 or more stars revolving around one main star like the planets in the solar system. Perhaps only single stars are the ones that have planets.

John Goodricke in the 18th century decided to make systematic observations of the star Algol in Perseus. He watched the variations in brightness for 11 nights in 1783 and noticed how Algol stayed dim for 7 hours once every 69 hours. By May 3 he knew he was seeing a dark star eclipse a bright one, and on May 12 he wrote to the Reverend Anthony Shepherd in Cambridge, England. From Cambridge the letter went by private stagecoach to London, and it was read before the Royal Society on May 15. (This was the Thursday night next to the full moon, the traditional monthly meeting of the Society. [Program AGE gives 13 days.] The members always arranged it that way because there were no street lights in Georgian London, and cutthroats and robbers preferred to work in the dark of the moon.) So within 9 days a scientific paper was written, and within 3 days, it was read to the scientific community. A speed record for the 18th century, and even for today!

Goodricke was 19 years old, handicapped and deaf, yet he was the first to discover a revolving, eclipsing binary star. Algol had long puzzled the ancients. The Arabic name Al-Ghul means the "changing ghoul." The Chinese Tseih-She means "piled-up corpses." The Hebrew Rosh-ha-Satan means "Satan's head." In the Greek constellation Perseus, Algol marks the severed Gorgon's head. In folklore it was the demon star, the blinking eye. Goodricke explained the mystery in terms of a dark orbiting star obeying the laws of Kepler and Newton. Unfortunately his own life was eclipsed by an early death at age 22.

The Königsberg astronomer Friedrich Bessel in the 19th century decided to analyze the motion of Sirius. He showed how Sirius was moving in a wavy line. He concluded that the star was being pulled from a straight line track by a companion star, and he calculated an orbit. He announced the existence of an "unseen companion" in 1844, but he was not believed.

The doubts continued until 1862, when Alvan Clark was commissioned to make a telescope for the University of Mississippi. The objective lens was to be 18½ inches in diameter, the largest in the world at the time. The glass blank was ground and polished, stored in a fireproof safe, and as an extra precaution, Clark rigged up a burglar alarm by his bed. When the lens was ready to be tested, Clark pointed it at Sirius. He saw what telescope makers dreaded—a double image. He was dismayed. He thought there was a fatal flaw in the lens. Actually the faint speck turned out to be the "invisible" star of Bessel, Sirius B. It has since proved to be a remarkable star, a new type of star, a white dwarf. The object is very compressed and tiny, yet it has a large mass and can swing giant star Sirius around in an orbit.

Robert Temple has argued that the Dogon tribe of Africa knew about the companion of Sirius long before the telescope, long before mindstep 3, and were told about the star by visitors from outer space. Subsequent research shows that the Dogon probably picked up the knowledge of Sirius B from earthly visitors, the 19th-century African missionaries.

A star is disappointing through the telescope. It shines brightly, but one sees no more detail than with the eye alone. William and John Herschel were obliged to shade their eyes when the 48-inch reflector was trained on Sirius because the brightness dazzled them, causing night blindness for half an hour afterwards. Their telescope increased the brightness by collecting a large amount of light, but the star would never show clearly as a disc because it was so far away. Nor was the 48-inch mirror anything like the accurate lens of Alvan Clark, so the Herschels had no hope of seeing Sirius B. Telescope astronomers were limited and frustrated by the enormous gulf that separated the earth from the stars. The secrets of all the stars were hidden in a fiery point of light; all the stars save one—the sun.

The sun was many things to many societies—a globe of burning rock, a disc of solid gold—but the notion of a solid sun dissolves in the telescope. The bright surface of the photosphere shimmers and explodes, high clouds float in a gaseous atmosphere, and flame-like flares burst with outpourings of energy from the deep interior. The sun is gaseous from the topmost fringe through to the core. Its outpouring of energy comes from nuclear fusion. Our planet's poison, nuclear waste, would be rich food for the sun if it could be taken there in a space shot.

During an eclipse the moon exactly covers the photosphere and the chromosphere comes into view, shining like a crimson ring. Then the outer atmosphere of the sun, the corona, shows shining pale green. The light comes from fluorescence as atoms of iron, neon, and other elements absorb a portion of the sunlight, hold it momentarily, and then send on the energy once again. Each fluorescing atom has a distinctive color. The light is absorbed and re-emitted with the trademark spectral line of the atom. The chemical composition of the sun can be figured out by spectroscopic analysis. The outer regions are 73 percent hydrogen, 25 percent helium, with the remaining 2 percent made up of all the other elements, the heavy elements of iron, tin, lead, carbon, nitrogen, oxygen, etc.

The corona and chromosphere are continually shaken by explosions from below. The photosphere is a seething inferno where mushroom clouds rise to the surface with a load of searing heat. The telescope shows millions of bright cells. Each one is like a giant H-bomb, spreading over 500 miles of territory. Sometimes the photosphere erupts with more than normal violence and the sun's material punctures its way through the chromosphere out into the corona. This glowing gas rushes out for thousands of miles, guided by magnetic lines of force, following a curved path up into the corona and back down to the photosphere. These gas streams begin and end at the edge of a sunspot, one of the dark blemishes first seen by Galileo. The spot itself is not a storm but a region of quiet, a cool patch isolated temporarily from the surrounding photosphere. Around the edges of the spot, however, the sun is in violent upheaval. Energy from the interior is welling up, causing explosions and ejection of material. Sometimes a portion of the photosphere is blown off completely, leaving the sun and reaching the planets as an ionized cloud, the solar wind.

A small part of the sun's energy falls on the earth to be absorbed by plants which provide organic food, to be preserved in the coal seams and oil deposits, and to supply the heat that warms the continents and oceans. The source of this energy cannot be seen. The view is blocked by the glowing layer of hydrogen in the photosphere. Below this level the sun becomes unobservable, and the nearest star reveals no more through the telescope than those other distant points of light.

Two thousand stars show up in the constellations on a clear

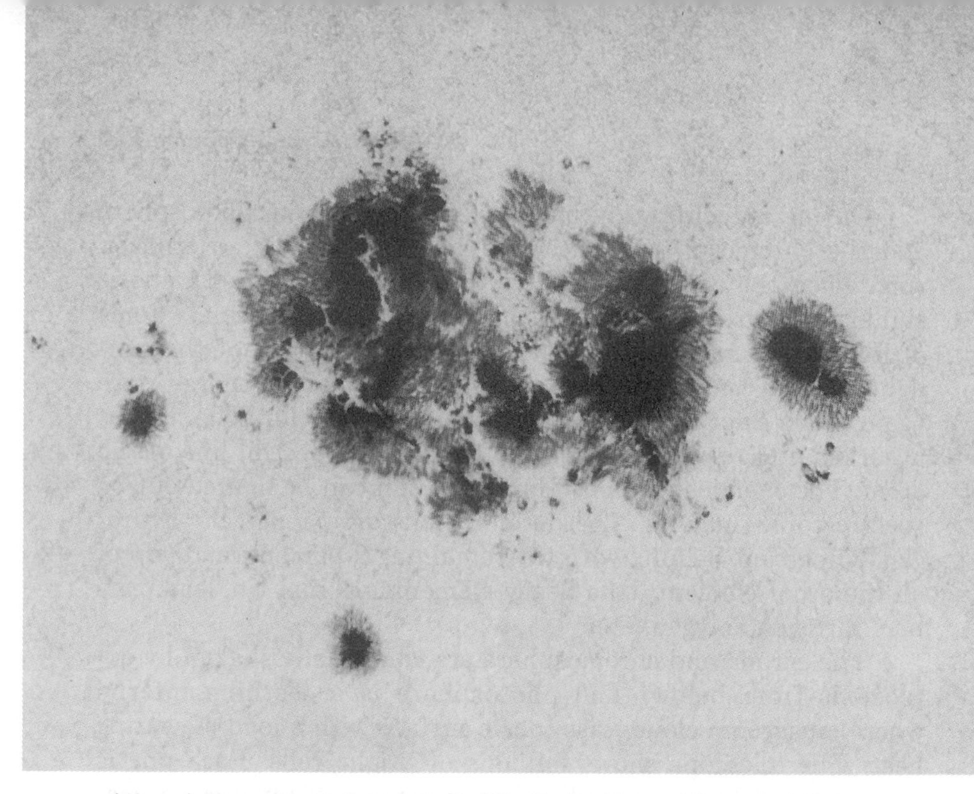

(Above) Sunspots and granules in the photosphere. (Opposite) Storms on the surface of the sun.

moonless night, 100 billion stars populate the sun's galaxy, and countless galaxies each have billions of stars, yet most of these stars fit into a simple 7-fold spectral mold. The temperatures of the photosphere surfaces range from 25,000 to 2,500° K (K for Kelvin, the absolute temperature measured as "Celsius plus 273"). If it were any cooler the star would not shine; if hotter, it would blow apart with the pressure of the outward-streaming energy. Stars, it seems, are never more than 20 times the diameter of the sun, and never smaller than one-tenth. Short-lived stars can be bigger, and white dwarfs can be smaller, but these are exceptions to the rule. An object much smaller than one-tenth of a solar diameter would be cold. It would be an icy planet.

The 7 spectral types are spelled out in letters—OBAFGKM—with O as the hottest and M as the coolest. The sun is a G star with a surface temperature of 5,800° K. Older stars of Population II tend to be half as bright, class by class, as the stars of Population I. The hottest stars are blue-white, the coolest dull red. The sun, after allowing for the orange hue given by earth-atmosphere pollution, is a pale yellow-green. When all the stars are plotted out on a diagram of brightness versus color

they form an orderly array known as the main sequence, stretching from the blue-white O stars to the red M's. It is a sequence in name only, though—stars do not move along this sequence as they develop. They move across it.

Some stars do not fit the sequence. Red stars like Betelgeuse are hundreds of times brighter than average. This puts them to the right of the main sequence, on what is called the red giant branch. Other stars are even brighter—the super giants. They have spectral types from B to M, which places them across the top of the diagram. White dwarfs, like the companion of Sirius, have the whitish color of A and F stars but are a hundred times fainter. This puts them below the main sequence. Most of these "misfits" were on the main sequence at some time or other.

At first astronomers thought the space between the stars was empty. It is not. The spiral arms are filled with a tenuous gas which has collected into clouds, and even between the clouds there is still rarified material. A faint fluorescence of a deep red hue comes from the clouds, showing that the gas is hydrogen. The glow can be traced with sensitive instruments for most of the length of the Milky Way.

One of the most impressive clouds is the great nebula in Orion, near the stars of the sword. There are 4 hot O stars at the center of the cloud, which appear to the naked eye as a single star, the center star of the sword. There are many other O and B stars scattered through the cloud, and it is the ultraviolet radiation from these stars that makes the gas fluoresce. Hydrogen, oxygen and nitrogen absorb a portion of the starlight, and send it on with a different color coded to the particular chemical.

No gas escapes notice if it is near the giant stars of class O or B, for the intense radiation tears the atoms apart, causing it to shine. Hydrogen within a distance of 200 light years from an O star is split into protons and electrons. The hydrogen becomes ionized and the O star is found at the center of a sphere of broken atoms. Occasionally a proton and electron recombine to form a hydrogen atom, and during the process light is emitted at the hydrogen wavelength. But the formation of the atom is only momentary; almost immediately it is ionized once again by the radiation from the central star. A giant star illuminates an enormous sphere of gas around it. Some of this material has been dredged up from the core of the star and expelled into space.

The Great Nebula in Orion.

Black dust clouds silhouetted in Serpens.

In 1904 the German astronomer Johannes Hartmann was puzzled by an absorption line produced by the element sodium which did not seem to belong to the star at all. The same line was found in the spectrum of several stars from type O through M. He concluded there must be a cloud of sodium vapor between those stars and the earth. The chemical cloud had shown up as disturbance in the light of the background stars, and by tracing the foreign lines he, and other astronomers, began to map out the extent of these clouds.

Interstellar clouds contain a large number of chemicals: calcium, sodium, potassium, iron and more. In fact any element capable of absorbing light when in gaseous form has shown up. The gas seems to be composed of the same mixture of elements that makes up the atmospheres of the stars. In addition there are chemical combinations, molecules of carbon plus nitrogen, carbon plus hydrogen, and oxygen plus hydrogen. The basic building blocks of organic molecules float in the spaces between the stars.

The average star-to-star distance in a spiral arm is 8 light years. Alpha Centauri, the star nearest to the sun, is 4.3 light years away, a little closer than the average, but not close enough to make a binary. In some regions of the galaxy there are hundreds of stars closely packed together with separations of about a light year. This is close enough for gravitational interaction, and the stars hold together as a group, a galactic cluster. Clusters are a multiple-multiple star, a single system, but the stars do not move in simple Keplerian orbits. The gravitational motion is more like a random jiggle than an orbit. Speeded up, the motions would look like bees in a hive.

The local galaxy contains thousands of these clusters. Telescopes reaching out show 400 within a distance of 10,000 light years, and then the curtain of black interstellar dust blocks the view. Ursa Major, the Big Dipper, is a cluster. The 7 stars (excluding the stars at either end) and many of the fainter ones in the constellation are a stable, related group. The stars are moving through space toward the constellation Draco and will survive as a cluster for millions of years to come. The stars of Taurus the bull are a cluster, and in half a million years the stars will merge into the constellation of Orion.

Ten degrees northwest of Taurus there is a small cluster

called the Pleiades, the "seven sisters" of Greek mythology. Normal eyesight shows 6 stars; the 7th star supposedly deserted the group in antiquity, moving over to become faint Alcor, companion to Mizar. Here the legend does not agree with the facts, because Alcor is a binary of Mizar and must have been moving along with the Big Dipper cluster since those stars first began to shine. Binoculars show many more "sisters" in the Pleiades, and the photographic count goes as high as 250.

If a cluster is farther away than the Pleiades it cannot be seen well with the naked eye. There is a hazy patch of light in the constellation of Cancer called since ancient times the "Beehive." Through binoculars the hazy glow becomes a mass of stars, a galactic cluster. Why did the ancients call it the Beehive? Could they somehow see the faint stars, or was the name a coincidence? We do not know.

These clusters contain the bright, but short-lived, O and B stars. They are always found in the plane of the galaxy, embedded in the spiral arms. If the stars in a galactic cluster form from a single large cloud of dust and gas, then more likely than not the stars are the same age. They are the "young" stars of Population I. A typical galactic cluster is 30 light years across and contains 500 members. After condensing out, the cluster moves along in the spiral arm in which it formed, taking up the general rotation of the galaxy. The group of stars is kept together by mutual gravity. There are disturbing forces, however, which will slowly break up the cluster. After billions of years the cluster will become elongated until finally the aging stars are scattered along the spiral arm.

When clusters form away from the spiral arms, their shape is different. The stars make a compact round cloud, a globular cluster. Usually these clusters are bigger than galactics, with diameters of 120 light years, and they contain thousands of stars. Viewed from earth, the points of light merge together at the center, and individual stars are lost in the glare. The stars are about one-quarter of a light year apart, closer than stars in the spiral arms. If planet earth were in a globular, there would be thousands of stars brighter than Sirius, and the constellations would shine down on the landscape as bright as the full moon.

Globular clusters are set above and below the spiral arms, not in them. More than a hundred globular clusters have been photographed through gaps in the dust clouds, and perhaps an equal number remain hidden. They form a spherical halo around the nucleus of the local galaxy—a cluster of clusters.

Astronomers think that all the stars in a particular globular cluster, like galactic clusters, condensed out of the dust at the same time. They are older than galactic cluster stars because the short-lived O and B stars seem to have burned out. Globular clusters belong to Population II, the dead population with little dust and where no new stars appear.

About 1 degree northwest of zeta Tauri, the left-hand tip of the horns of the bull, there is a faint patch of light shaped vaguely like a crab. Photographs taken several years apart show that the nebula is a cloud of gas which is expanding. Working backwards in time, astronomers find that the cloud was a single point about 900 years ago. Something must have happened between 1000 and 1100 A.D.

At that time Europe was bogged in the dogma of mindstep 2: changes in the cosmos did not occur, period. But North American Indians, still in mindstep 1, painted a new star on their rock shelters, near the crescent moon. Over in China, astrologers scrutinized the heavens for unusual happenings. The observations were needed for preparation of imperial horoscopes for the various emperors. To-To wrote in the *History of the Sung Dynasty:* "In the first year of the period Chih-ho, on the day Chi-chou of the fifth moon, a guest star appeared near Tien-Kuan." Now Tien-Kuan is near zeta Tauri, and the Chinese date works out to be July 4, 1054 A.D., by the Old Style calendar. The Chinese had seen the outburst of light, the enormous explosion which caused the Crab nebula.

The guest star was visible by day as well as by night. Conclusion: the guest star was extremely bright, brighter than Venus at magnitude −4. It was a supernova. Yang Wei-te, chief astrologer of the Sung dynasty, forecast good times ahead because the star was the lucky color of the emperor, yellow. Later, in February 1055, the Liao dynasty astrologer predicted that the Peking emperor would die because the color had turned to red. The emperor did indeed die within 5 months, but bad omens aside, there are scientific inferences: the spectral change from yellow to red was due to the cooling of a superheated fireball, the explosion debris of a self-destructing star.

Globular cluster in Canes Venatici.

The Crab nebula in Taurus.

As mindstep 3 came to a close the telescope began to fail the needs of astronomers who then turned to mathematical exploration. Modern physics was used to deduce the life histories of stars and the nature of their invisible interiors.

A star like the sun condenses from a cloud of dust and gas—those dark globules photographed against the background of a bright nebula. Gravity draws the dust and gas together and interstellar matter falls in toward the center. Energy is released by the falling material, the cloud becomes warm, then hot, and shines as a dull red proto-star. Rare photographs have been taken of proto-stars in the Orion nebula at the instant they began to shine. The proto-star contracts for the next 30 million years, which is a short span in the life of a star. Contraction energy is released, the heat builds up, and the spectrum changes from red to orange and yellow. The color-temperature changes take the star from class M through K to G, and it becomes fainter. This is stage 1—the star drops down toward the main sequence.

With a central temperature of 10 million degrees K, the star begins to fuse hydrogen into helium. The nuclear reaction is steady (and safe!) and continues until most of the hydrogen at the center of the star is converted. The sun has been on the main sequence, so the mathematics indicates, for 5 billion years, and will continue to burn steadily for another 5 billion. All that time, gravitation will be balanced by gas pressure and by the force of the escaping radiation.

If the original globule of dust is small, the star takes up a position at the lower end of the main sequence. It becomes a faint cool star, class M. Larger globules produce stars at the top of the main sequence, the brilliant O, B and A stars. These stars are hotter than the sun, and when the central temperature goes higher than 20 million degrees K, the nuclear reactions become violent. Carbon acts as a catalyst, speeding up the conversion of hydrogen into helium. The stars burn more rapidly, shine brighter and have a lifetime much shorter than does the sun. According to the math, O stars stay on the main sequence for only 10 million years. That is why O stars are rare—they burn out quickly.

This hydrogen fusion process on the main sequence is stage 2. Helium, the product of the fusion, slowly builds up as a dead core, clogging the furnace. At first no more than a mile across, the core spreads until it takes up 10 percent of the star's mass.

There is no stirring action to break up the core, and although the center is hot there is no hydrogen left to burn. After 10 billion years for a sun-size star, hydrogen fusion ceases and the star leaves the main sequence.

In stage 3 the sun will become a red giant. When no more fusion energy is released in the core, when there is no outflow of radiation to hold it up, the core begins to contract, as the original dust globule did some 10 billion years previously. But the core is more compacted than the dust cloud and releases much more energy. The heat pulse makes the outer layers of the star expand. The star becomes bloated, increasing in size by 100-fold. Five billion years from now the sun will reach this stage. It will vaporize the planet Mercury, melt Venus and singe the earth. From earth the sun will swell into a red ball of fire, extending halfway across the sky as the photosphere cools slightly, changing from yellow-green to red. At this stage the earth's oceans will boil off, the atmosphere will leak away and the rocks of the surface will melt. Planet earth will be a lifeless inferno, returning to the proto-planet conditions of its beginning.

In stage 4 core melt-down occurs. The core overheats, reaching temperatures of 100 million degrees or more, and the star is then at the topmost part of the red giant branch with maximum diameter and brightness. Now helium nuclei fuse to make carbon and all the other chemical elements. The outer layers contract, the gases become more compressed, and the rise in temperature changes the color from red to yellow.

Mathematics begins to fail in this stage of development because the shrinking does not go smoothly, and rhythmic oscillations develop which defy calculations. The star recrosses the main sequence, but through the telescope it can be distinguished from a normal main sequence star by the pulsations. The variations in brightness occur quite rapidly within a period of a few hours. There are lots of these rapid variables in the older star clusters.

Stage 4 is the most important one for planets and the chemistry of life. The core temperature in its melt-down approaches temperatures of 5 billion degrees. Multitudes of fusion reactions take place. All the heavy elements of the chemical table are produced at this stage. The star is a factory making quantities of the 91 elements between helium and uranium—oxygen for atmospheres, iron for blood and carbon for vegetation. The precious metals, gold and silver, rare earths for TV screens, silicon for the chips, are all created by fusion in stage 4.

These elements are dispersed into the dust clouds of space during stage 5. At that time the heavy elements in the carbon core clog and stop the intense nuclear power plant. Only a thin coating of hydrogen remains around the outside of the core, and the supply of hydrogen fuel is almost exhausted. Enormous forces of gravity take over, squeezing the star down to the size of a planet. It is small, faint in light output, and very dense. It has become a white dwarf. There is no space between the electrons. The core is electron-degenerate. A sugar cube taken from a white dwarf would weigh several tons. Placed on a table it would crash right through. According to the calculations, electron-degenerate stars are unstable. The photosphere will blow off into space. This may be the cause of supernova explosions.

Degenerate stars may also be the cause of the so-called planetary nebulae. When these hazy discs of light were first seen in the early telescopes they were mistaken for planets. But they didn't move and had to be far outside the solar system. They are gas spheres and the star at the center is blowing off material. The degenerate core is slimming down to the white dwarf stage. Calculations show there is a process of convective dredging going on. Circulating gases in the outer layers are dredging up the heavy elements from the core which are then ejected into the surrounding space. Time has corrected the original misnomer. They looked like planets but they were not, and now they are found to be enormous factories spewing out just those special chemicals needed to make future planetary systems.

In stage 6 the contraction goes one stage further—to neutron-degeneracy. The star collapses to the size of an asteroid, a few miles in diameter at the most. There is no space between the neutrons, the nuclei of the atoms. The core is now a neutron star.

Stage 7, if there is such a step, cannot be followed in the telescope because the star cools down to a dull, nonluminous cinder—a black dwarf. Or, if the contraction goes beyond a certain critical limit set by relativity theory, stage 7 could result in a super-massive object from which no light waves could escape—a black hole.

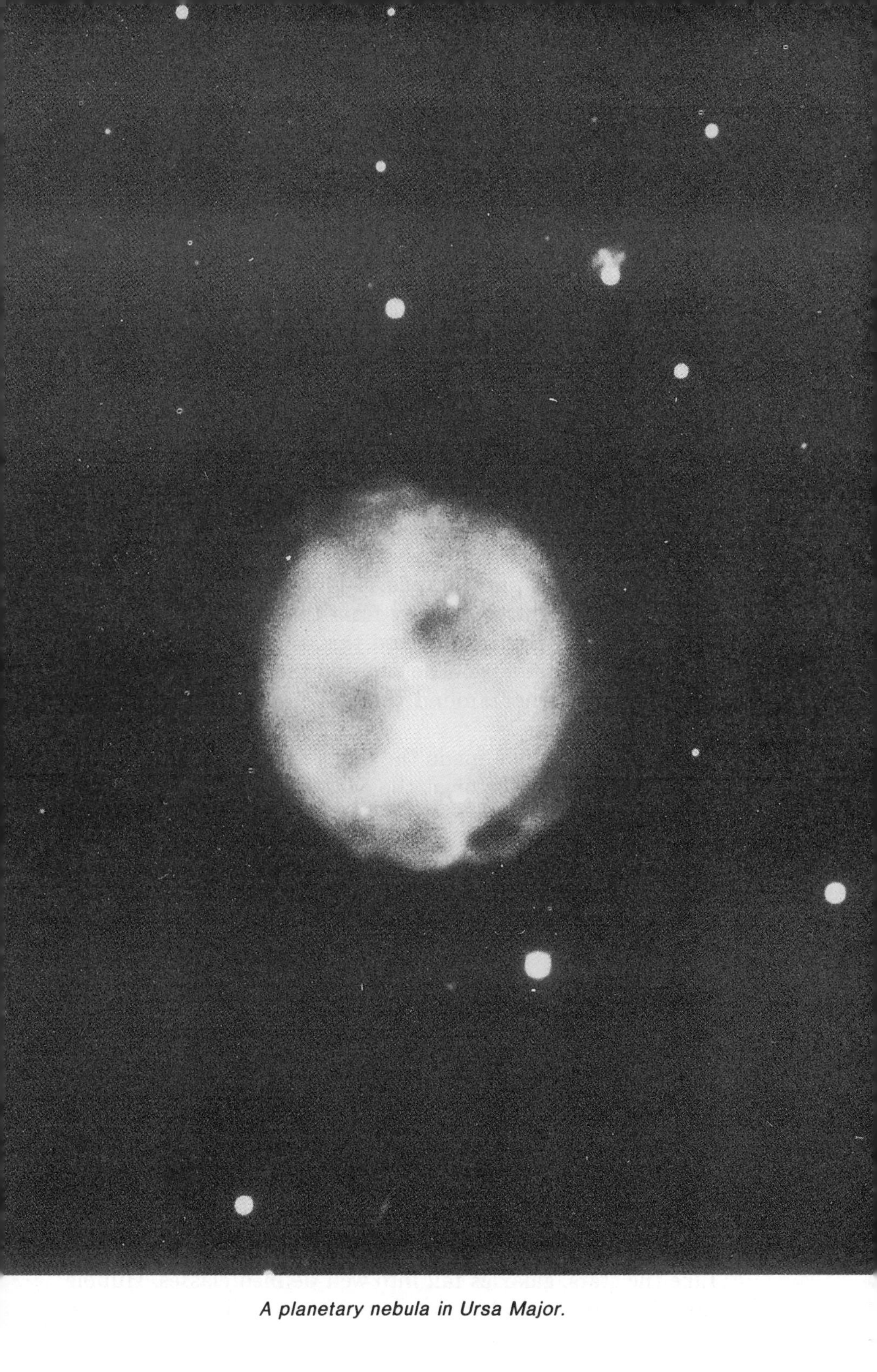

A planetary nebula in Ursa Major.

Mindstep 3, running its course through the post-Renaissance centuries, was the most difficult adjustment for humans to make in assessing their place in the cosmos. In a way they were told more about the universe than they ever cared to know. The step was an informational revolution in which a large, safe home, fixed at the center of an ordered universe, shrank down to an infinitesimal cosmic mote. Copernicus downgraded the earth by making it a moving planet with the sun as the hub. Galileo's telescope showed that the earth was one of the smaller planets. His little telescope resolved the Milky Way into stars, and the sun was one among a myriad of stars. Up until the early 1900s astronomers believed that the sun was at the center of the Milky Way galaxy, and that—yes!—the Milky Way was the entire cosmos. But Harlow Shapley toppled that last idea of a rightful place. He showed by photographs of globular clusters that the center was 30,000 light years away from the sun in the direction of the constellation Sagittarius, and that the sun was in an arm of stars near the outer edge. The sun and the neighborhood stars make a circular journey around the nucleus of the galaxy once every 230 million years.

Then Edwin Hubble made the first dramatic discovery with the "new" 100-inch reflector at the Mt. Wilson observatory: the hazy patch of light in Andromeda was a spiral galaxy, a Milky Way like our own, with 100 billion stars. Other smudges on the photographs were other galaxies, endless island universes in all directions of space. The collective mind was numb. It had to adjust to a revolution in assessing its place, its existence and its future.

Long-exposure photographs taken with the 200-inch telescope on Mt. Palomar penetrated to a magnitude of +23, about 15 million times fainter than the faintest star visible to the naked eye. These photographs showed 5,000 galaxies in a square degree. With this packing density a period on the page of a book (.) would cover up 10 whole distant galaxies, so prodigious are their numbers. At the limit of its range the 200-inch telescope could reach 1 billion galaxies. Counting galaxies like pennies, "1, 2, 3, . . . ," it would take 100 years to reach a billion, and that is counting steadily 7 days a week, 8 hours a day.

Like the stars, galaxies fall into well-defined classes. Hubble recognized 3 types: spirals, barred spirals and elliptical galaxies.

The Andromeda galaxy.

A field strewn with spiral and elliptical galaxies.

Then he divided the 3 types into the sequence shown in the diagram.

Elliptical galaxies take their name from the shape. They are like huge globular clusters that have been distorted from the spherical shape, and Hubble's number gives a measure of the elliptical flattening. E0 galaxies are circular in outline while E7 are extremely flattened. Ellipticals resemble globular clusters in another way—they contain Population II stars, the old first-generation stars. No young stars are seen in the telescope, and presumably no new stars are forming at the present time. There is not much star-forming dust in ellipticals, and the stars seem to belong to a closed system, doomed to a black-dwarf extinction.

Spirals have a bright nucleus at the center of a disc of stars. The nucleus is composed of Population II stars, and the disc, Population I. The disc of a spiral is filled with thick clouds of dust from which young stars are forming. Hubble's sequence has the nucleus of the spiral become smaller and smaller until at type Sd it has almost disappeared.

Galaxies vary in size. There are giants and dwarfs in each class. Ellipticals in general are smaller than spirals. They are 10,000 light years across and 3,000 light years thick. The average Sb spiral is 50,000 light years in diameter and about 8,000 thick. The Sa and Sc spirals are somewhat smaller, being about 30,000 light years across on the average.

With the 100-inch telescope astronomers could get to a range of 5 billion light years. With the 200-inch and further special improvements, the range increased to 15 billion light years, but at those far frontiers reached by instruments, astronomers were dealing with galaxies and clusters of galaxies that never had been seen with the eye. The light, enfeebled by millions of years

The classification of galaxies.

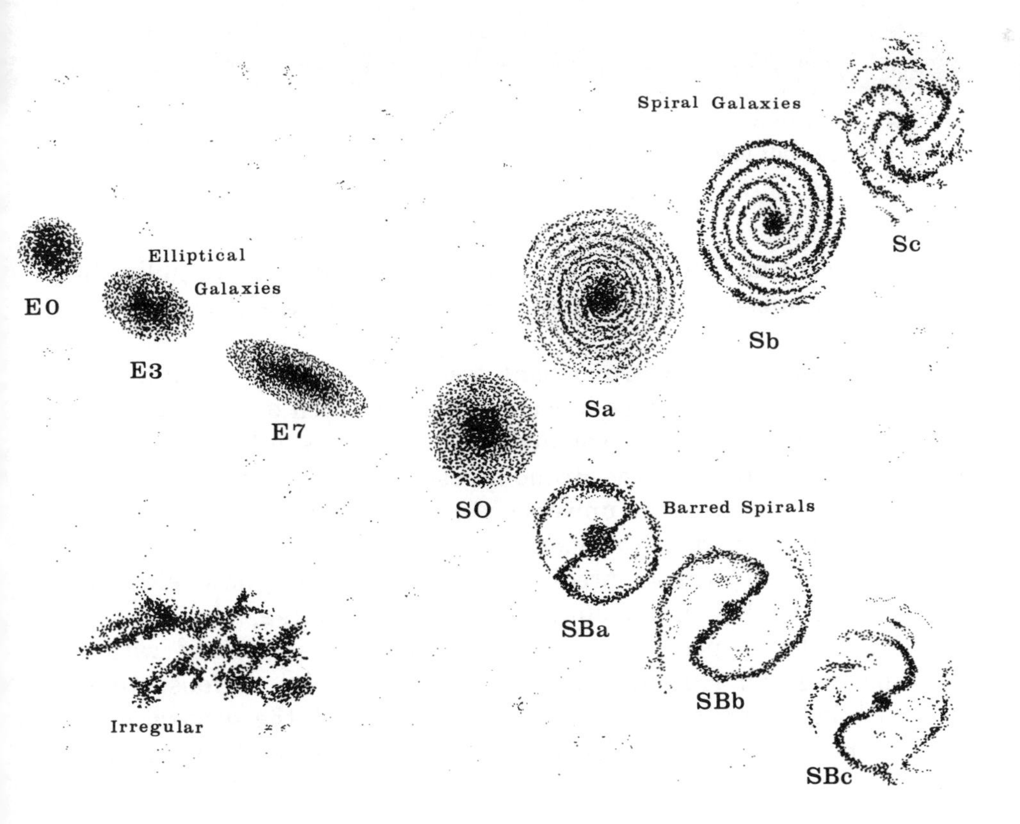

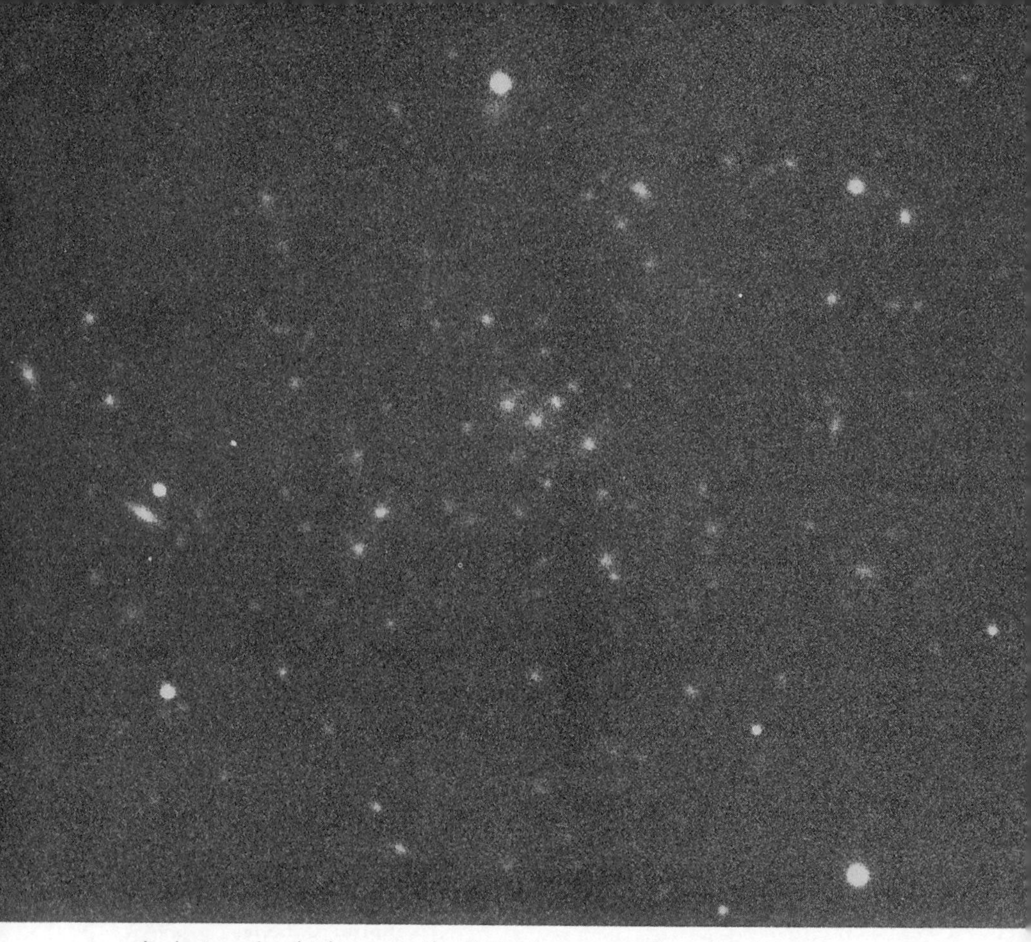

A cluster of galaxies near the limit of the earth-based telescope, more than 4 billion light years away in Pisces.

of travel time, was only strong enough to produce a sprinkling of silver grains on a photographic film, or to generate electronic signals.

Was Giordano Bruno right, was the universe truly infinite? If there was a real edge it was beyond the reach of the telescope. Galaxies and clusters extended in all directions of looking. There was no way to spike the Bruno heresy with simple galaxy counts.

Not only was the universe seemingly endless, it looked as if it was blowing apart at enormous speeds. Albert Einstein's general relativity theory predicted, and Edwin Hubble's telescope revealed, this expansion. Galaxies, of course, showed the smeared-out spectrum of all the component stars, and with a spectrograph Hubble compared the light from various atoms with light from similar atoms on the earth. All the spectral lines of the galaxies were displaced to the red end of the spectrum, and the redshift proved to be greater for the more distant galaxies. The shift in frequency was likened to the change in pitch of a note from a moving object, the so-called Doppler effect. (The whistle of a train moving away is lower in pitch than that of one approaching.) But for galaxies the effect was extreme. Some of the distant galaxies were clocked by their spectral lines at a speed of half the speed of light.

There was a connection between speed and distance, the speed increasing by 10 miles per second for every step of a million light years into space. The relentless increase with distance could not go on indefinitely. The theory of relativity predicted a speed limit in the universe of 186,000 miles per second, the speed of light. Nothing could crash the "light barrier" because theory predicted that all masses become infinite at that speed. Astronomers wondered what would happen at a distance of 18.6 billion light years. Hubble's law predicted a recession velocity of "c," the speed of light. Surely that would be the absolute edge of everything.

Physicist George Gamow reversed the argument. He showed how the speed-distance connection was exactly right for all the galaxies to have been compressed together about 18 billion years ago. Marching back through time in billion-league boots, all the galaxies, all the stars, the planets, everything merged at that ground-zero date into a super-compressed fireball. According to the theory, at that date, in the twinkling of a microsecond, a white hole suddenly appeared in space. Hydrogen condensed out and 25 percent of the mass was fused into helium. At a temperature of 3,000 degrees K the expanding gases had cooled sufficiently to be transparent, and stars, galaxies and clusters of galaxies condensed from the material. Things continued to move out from the center of the explosion as cosmic evolution unfolded—Population II, I, multiple stars, single stars and planets. Theoretically the universe was destined to end either in ice—as galaxies expanded into black space, consumed the supply of nuclear energy and cooled down—or in fire, where the expansion slows, the motion reverses and the big bang universe turns into the big crunch. After that, a new white hole appears in space, made up of all that ever was, and containing all that ever will be.

Looking outward into space was also looking backward in time. Light from the Andromeda galaxy arrives at the earth after a journey of 2.5 million years. Photographs show the galaxy as it was 2.5 million years ago. Those smudges on the film at 5 billion light years' distance are shown as they were at the time planet earth was condensing from the solar nebula. Distance automatically gave astronomers a backward-looking time machine. In truth, all those galaxies could have succumbed long ago, yet astronomers would have no way of knowing.

By the mid-20th century, competing cosmologies had appeared. The big bang was the most highly favored, at least by Western astronomers, because it was the most comfortable fit with ancient mindstep 2. That step was the age of order, of dogma, and coincided with the emergence of the world's main monotheistic religions. It was as though the Book of Genesis had been rewritten with a mathematical appendix: "In the beginning there was created in a fraction of a microsecond a giant, primeval fireball from which everything was made, according to the laws of physics . . ."

British astronomer Fred Hoyle brought out the "steady state" theory. He said there was a term missing from Einstein's general relativity equations, and so Hoyle added a creation tensor. With this change in theory, the universe did not begin all at once in a primeval fireball. Hydrogen ions and atoms were continually showing up between the galaxies as the system expanded, and the density of the cosmos was magically adjusted always to stay the same. Hoyle's theory got a lot of support, and much of the evidence was on his side of the argument at the time. According to Hoyle, the universe would always look the same, always be the same, as new galaxies replaced the old from the clouds of continually created hydrogen. The cosmos would grow indefinitely like an ever-unfolding flower.

Opposition to Hoyle came from the pulpit of Canterbury Cathedral. The then archbishop denounced the theory on a theological fine point: God could not be continually creating hydrogen in every far-flung corner of the universe. To say there was a continuous creation tensor in the Einstein equation was to deny the existence of God.

In 1960 and 1962 I published a small contribution to the cosmological debate. Working from the Einstein metric I suggested that the universe was balanced by a negative pressure, static and nonexpanding, and that the redshift was due to gravitational effects. The origin of time was at minus infinity, which implied no instant of creation. The cosmos always was, is now, and always will be. It is, mathematically speaking, eternal. For it to work, matter and energy must be used and reused. Decaying stars must be recycled into the dust clouds. Cosmic energy release, absorption and release again must be 100 percent efficient when averaged over the whole. As one more theory on the scene, the "eternal universe" was by far the least successful. A lead balloon thumped in the rarified air of cosmology. A few astrophysicists asked for a reprint (in those days there was no such thing as a photocopier), but they showed zero interest in exploring the mathematics of a model where the galaxies were not flying apart and in which there was no creation.

However, 20 years later, in April 1982, Dr. Paul Savet, former director of research at Grumman Aerospace, wrote:

> Prof. Hawkins is correct therefore in saying, or implying, that a static universe with a prevailing gravitational field is equivalent to a quadratic expansion. The Hubble expansion is due on the other hand to a density gradient and corresponding space-curvature.
>
> In textbooks on General Relativity the derivation of the above is based on the ds^2 of space, which includes for instance the Schwarzschild coefficients of the metric. Yet the field equations involve not only the Ricci tensor R_{ij} augmented by $-\frac{1}{2}Rg_{ij}$, but another second rank, in principle undetermined and therefore arbitrary, tensor. This is a weakness recognized by Einstein and others which could never be remedied. The Hawkins approach is very elegant. It proceeds by direct derivation[*] and yields a correct interpretation of the redshift with no arbitrary elements in extra-gravity space curvature. . . .

As mindstep 3 reached its full development, the position of earthlings changed in an irreversible way from order to expanding chaos, from a place at the center to no place at all, from a future controlled by divine will to sure prediction of disaster by either fire or ice. There was a succession of reappraisals, revolutions in thinking. The sun as anchor point was destroyed, to be replaced by the nucleus of the Milky Way galaxy. Then this anchor point was found to be no different from countless other galaxies, and all were moving.

*G. S. Hawkins, "Expansion of the Universe," *Nature,* 194 (1962):563–64; "The Nature of the Redshift," *Il Nuovo Cimento,* 23 (1962):1021–27.

Back in the days of flat earth, and even with a Columbus-size globe, there was a respectable place for earthlings. There was a human dimension to the then known cosmos. To be sure, the oceans, mountains, deserts, the vastness of the planet, were dangers, challenges, but earth was there to be cataloged and conquered, used and nurtured. The sun, moon, stars and other planets were all part of the simple picture.

As the impact of the 3rd mindstep unfolded, the place of humans shrank to nothing. To passive, earthbound observers looking out, the physical universe became impossible to comprehend. Nothing in its heritage had prepared the world for such a shock. Seen through a glass darkly at the limits of the telescope, the edge of the universe, filled with explosions and runaway galaxies, was as threatening to the mid-20th-century earth as that awful edge of the Atlantic Ocean had been to Columbus's mutinous crew.

As it happened, mindstep 3 had run its course. Continuation in the role of nonparticipating audience would have produced more revolutionary discoveries, more demeaning facts, but no change of status. Continuing in 3, Sapiens the "wise" would turn inward to the frail ecology of the earth and try to adjust to a minimal, or zero, role in the greater cosmos. Without a further step, Homo sapiens, the only species known to be aware of the cosmos, to have higher intelligence, would become isolated, marooned on a speck of rock awaiting the red-giant phase of the sun. Legendary Gilgamesh was told to trim his ambitions as a mortal—eat, love, reproduce, die—that was the alpha and omega for man in mindstep 1. And the Copernican mindstep at its furthest reach would be no more profound than Gilgamesh's. But a new perspective, even more bewildering and challenging, came along, and we live it today. . . .

TWELVE
The Pod Door Opens—Mindstep 4

It is difficult to fix the precise date of a mindstep, that instant when a new cosmic idea is certain to survive on earth. Precursors appear, show flittingly and die without connecting to the main thrust. We identify them only by hindsight, not because they were the root cause of the change of viewpoint. Then again, a mindstep starts quietly, almost unnoticed, like the first stirrings on the ocean surface that are later to develop into a great, crashing wave. For centuries Ptolemy's ordered system was known to only a few. Copernicus's revolutionary idea did not get into the textbooks of his time for almost 50 years. People can live and die untouched by a mindstep, left behind like flotsam as the change of perspective gathers strength beneath them and the rising crest moves on into the future.

The old perspective has once again been shattered. We are no longer a passive audience gazing in wonder at the universe, be it a dome or infinitely extending space. We are no longer encapsulated, separated from the cosmic environment. The pod door has opened and human beings, for better or worse, have taken the first steps into the unknown of space. The required mental adjustment is mindstep 4. In all previous mindsteps—0, 1, 2 and 3—humans looked at, up or out at the universe. From this point on it is a domain in which to move, to travel, to touch and live in. The new idea began to spread irreversibly sometime during the last 50 or 100 years, and by the mid-20th century, leaving the earth and earth's atmosphere had become a reality. Life on earth could not be the same again.

When did mindstep 4 begin? Was it in 1865 when Jules Verne wrote his science fiction novel *From Earth to Moon*? Perhaps, but perhaps not. He did have the foresight to name Cape Canaveral as the launch site, but the story had all the elements of an earthbound adventure. The capsule itself was a miniature earth-environment with velvet cushions, cigars and even a homely dog. Nor was the story remotely realistic. The occupants were fired from a cannon, like performers in a circus ring. To my mind it was an earth-bound romance in a celestial setting. A precursor, yes, but no more continuous with the genuine mindstep than the space-travel dreaming of Kepler, or the fiery wheel of the prophet Ezekiel.

Was it in 1895 when the shy, modest math teacher Konstantin E. Tsiolkovski wrote about leaving mother earth with liquid-fueled rockets? Hardly so, because what Tsiolkovski wrote lay entirely forgotten until the Soviets brought it out in the 1960s as a claim for Russian space precedence. Was it in 1898 when British author H. G. Wells published *War of the Worlds,* a story of planet earth being invaded by Martians, or his fairly realistic journey described a year later in *The First Men in the Moon*? To be sure, what he wrote was avidly read, but it was mostly discounted as just science fiction. As a boy I was told never to mention this story in the presence of my grandmother. Even though she was a Kentish Wells and a close cousin of "H. G.," she and her family would never recognize the famous author as a relative, let alone listen to his "strange ideas." But there was a tremendous impact in what he wrote. Orson Welles (no relation) dramatized this Martian invasion on CBS radio in October 1938, and many listeners by that time accepted everything as true. Unlike Tsiolkovski's work, the Wellsian was strong and continuous and came through loud and clear.

Other writers took up the theme of interplanetary travel at the engineering, practical design level. The Germans, Herman Ganswindt in 1891 and Herman Oberth in 1923, were important contributors.

Or did the pod door open just a crack on October 19, 1899, at the top of a cherry tree on Maple Hill in Worchester, Massachusetts? Young Robert H. Goddard was in the branches, trimming the top of the tree, when he had what can only be described as a vision. He was 17 at the time, but whatever it was that burst into his mind stayed with him and drove him for the rest of his life. He never shared the details with anyone, but it seems he saw the ladder as only one small step, and above the topmost branches of the tree, beyond the blue of the autumn sky, he saw himself traveling by rocket to the moon and stars. Each succeeding year on October 19th he took his family to the foot of the tree for a celebration. It was almost a religious rite. He called it "Anniversary Day." Twenty-seven years later, after many failures, he made the prototype of all space vehicles, the liquid-fuel rocket. On March 16, 1926, this new creature of human prowess took off from a frozen field near his home and ascended 184 feet into the air. Goddard knew it would have done better in the vacuum of space. The earth's surface was an anathema to Goddard. The cosmos was his goal.

Not so at Peenemünde. There the development of the V-2 rocket (Vergeltungswaffe, Revenge Weapon 2) was intent on a terrestrial goal, England. I remember looking out of the chemistry laboratory window overlooking the North Sea as a schoolboy in 1944 and seeing the slowly rising rocket trails 200 miles away over Holland. Ten minutes later the preprogrammed warhead turned earthward and exploded in the marshes behind the town. After the warhead hit I heard the loud rumble of the rocket approaching—traveling faster than the speed of sound, the bomb came down and the sound came after.

Yet out of Peenemünde's revenge came the postwar international space programs. The chief V-2 scientist, Wernher von Braun, was like Robert H. Goddard. He had his sights on outer space all the while. He escaped from Peenemünde and asked for asylum in the United States, where he was the brains behind the National Aeronautics and Space Agency, NASA. Other German rocket experts were captured by the Soviets, taken to a site near Moscow and had their brains picked to provide the know-how for the Russian space program. Later, other countries also proved their capabilities, and Indian, British and French satellites went up. China founded her Academy of Sciences in 1949, and "starting from scratch," as the official information booklet says, "she launched her first satellite in 1970, and in December 1975 she successfully launched and accurately recovered a satellite, becoming the third country in the world to do so." China recouped, as it were, her slow acceptance of mindstep 2 with an early acceptance of 4.

Another irreversible hold that this 4th step gained on the earth's collective mind occurred with the launching of Sputnik I on October 4, 1957. Media-wise the impact was as great as the bogus dramatization of *War of the Worlds,* but now the coverage was not limited to the New York area—it was worldwide.

For bureaucratic reasons the scientific U.S. plans for the conquest of space were separated from U.S. military capabilities. With a woefully limited budget and the exclusion of von Braun's team then working at Huntsville, Alabama, a small 6-inch sphere, Vanguard, was scheduled to be launched sometime in 1957—known earlier as the "minimum orbital unmanned satellite experiment" (MOUSE), with emphasis on the minimal. Vanguard never made it on schedule and was soon swamped by later developments when von Braun joined NASA.

I was a staff member of the Harvard-Smithsonian Observatories at the time of Sputnik, and hearing the news on a car radio I drove immediately to the Vanguard tracking center, still only half built in the botany building of Harvard University. The place was bedlam. Dozens of press reporters crowded in demanding to know everything there was to know about orbits, space travel and the outer cosmos. Their first reaction was to denigrate, to disbelieve the Soviet achievement. "It can't be called a moon," they said. But Fred Whipple spoke for the astronomers and said: "Gentlemen, if it makes one complete orbit, the earth has a second satellite." Sputnik did. In colloquial Russian the word means "fellow traveler." Radio signals were picked up by a ham operator in New Jersey as the shiny 20-inch globe made its 2nd pass over the United States. The press took Whipple's evaluation and headlines flashed in the world's many languages.

At dawn we got the first pictures. We used a makeshift tracking system—a 10-inch camera timed with a manual shutter. The satellite's rocket casing shone in the sunlight of space and moved silently across the starry background, powered by earth's gravity. Period: 1 hour 20 minutes. Speed: 18,000 mph plus. Planet earth now had a 2nd moon. "I can't believe it," said one technician. "Goodbye Copernicus!" said another.

Now, only a quarter of a century later, thousands of space

craft circle the earth, dozens circle the sun, some are on their way to the stars, probes have reached or are heading for every planet except Pluto, and astronauts have trod the mountains and plains of the moon.

Our new perspective on the cosmos has already produced surprises, and there are surely more to come. Astronomers stuck at the bottom of an ocean of air could never have found out the true facts of the physical universe with their telescopes. Many of their hard-won "discoveries" were proved to be wrong, and their theories tumbled to the stark forces of reality. From the beginning NASA published its maps and pictures right side up. The old pictures in the textbooks now have to be uninverted to agree with what the space probes see.

Take Mercury, for example. Glued to the eyepiece of the telescope, astronomers mapped dark markings for over a hundred years and said Mercury turns on its axis once every 88 days so as to always face the sun. Wrong. The planet's period is 58.6 days. Mercury is locked to the sun in a strange resonance: at one perihelion passage the "front" faces the sun, and at the next perihelion it is the "back." This resonance will continue indefinitely because Mercury is slightly torpedo-shaped, and the sun pulls first at one end and then the other. This resonance tricked the unsuspecting astronomers, who happened to get their best view at one particular point of the 3-pointed star that Mercury makes in the zodiac. Three synodic periods of 116 days equal 6 rotations to within a few days, and so the dusky markings looked the same. Because of this the telescope astronomers (wrongly) said the planet always faces the sun.

Nor were the dusky markings anything like the true picture. Close-up photography of Mercury by space probes reveals a wealth of detail—craters, ridges, canyons and mysterious wrinkles. At first glance Mercury is the twin of the moon, except there are more craters and fewer plains. No one expected to see so many craters there, but like the moon Mercury must have been bombarded by huge meteorites in the early, formative days. It is a very dense planet, the heaviest of the planets for its size. The only explanation for the heaviness is the core—Mercury must have an iron-nickel core like the earth. Spacecraft detected a strong magnetic field enveloping the planet, which is another indicator of the existence of the core.

Mercury is cratered like the moon.

Mercury has no atmosphere except for the faintest trace of the inert gas helium, and so there can be no life. A prerequisite for all life as we know it is an atmosphere with which to interact, be it carbon dioxide or oxygen. Nor are the temperatures on the surface bio-conducive. Under the noonday sun it reaches 700° K, hot enough to melt lead and tin.

What of the cracks and wrinkles? Those markings are huge hills hundreds of miles long and thousands of feet high. They run through the plains, clear through the craters. The theory is that Mercury formed even hotter than it is now and slowly cooled over a period of billions of years. In the cooling process the surface wrinkled like a wizened fruit.

Venus is another inferno—the least hospitable planet in the entire solar system. The surface is 50° hotter than Mercury, atmospheric pressure is 90 times greater than the earth's (sufficient to crush a scuba-diver's watch), the air is mostly suffocating carbon dioxide, and the clouds are droplets of sulphuric acid—one of the most corrosive liquids known.

All this heat comes from the sun. Thermal rays penetrate the clouds and the heat is trapped by the carbon dioxide in the air. The result is a furnace with winds tearing over the surface at hundreds of miles per hour. Down below the clouds it is about as bright as an overcast day on earth, but on Venus the clouds are orange-colored, and the overcast is 30 miles high and shapeless.

It was this continuous cloud cover that thwarted earth-based astronomers. Radar beams have now been used to penetrate the clouds and reveal the surface. Because of the intense heat there is no water, but there are raised continents and depressions as if planet earth had been stripped of its oceans, seas and rivers. Even though the atmosphere is thick and might be expected to protect the surface from meteorite bombardment, there are a few craters caused by impacts. There is also a large oval caldera about 35 by 50 miles across, which is thought to be volcanic. Lava has flowed out to make a raised shield about 400 miles across.

Soviet scientists have managed to land instruments on the surface. The first capsule dropped by parachute was instantly burned up. Later models were refrigerated before release from orbit and protected by thick layers of insulation. They lasted for 3 hours and sent back pictures of the surface and data on the atmosphere. One of them landed near the large caldera and confirmed, as expected, that the rocks were of volcanic composition. Other capsules got pictures of distant hills, showing that the air beneath the murky sulphuric clouds was fairly clear. There was no life at all on that life-destroying planet.

But if Venus is an inferno today, it was worse in the past. Space scientists conclude that the planet was hotter and more active volcanically in its previous history. It was those violent upheavals that put the enormous quantities of carbon dioxide into the atmosphere, and ejected the sulphur which made the acid.

As with Mercury the rotation rate was a surprise. Venus rotates opposite in direction to the earth and other planets. And it is slow—249.9 days. Why this should be no one knows. As a further strange coincidence this slow rotation synchronizes with the 5-pointed star of Venus's manifestations (12 rotations equal 5 synodic periods). Either this is entirely due to chance, or there is a weak gravitational resonance between earth and Venus. Whatever the reason, this synchronization causes difficulties in earth-based studies. The radar probing was done when Venus was nearest the earth at successive inferior conjunctions and when the same face of the planet was presented to the earth. Only when a radar was put in orbit around Venus did we see the whole planet.

Could Venus ever cool down and develop oceans, plants and life like the earth's? Probably not. Venus has several strikes against it. First: it will stay hot as long as the sun stays hot, and the sun is due to get hotter in the red-giant phase. Second: there may not be enough hydrogen left on Venus to combine with oxygen and form water. Third: no one knows of any natural chemical process that could remove all that lethal sulphuric acid.

Earth is different. Even if it had not developed life it would still be the jewel of the solar system. Astronauts say that seeing it from space was an unforgettable experience, like Goddard's Anniversary Day but in reverse. International boundaries disappeared on that color-patchwork globe. Sovereignty melted away in the striking blue of the atmospheric sheath. Life was throbbing on that delicate curved surface beneath the pearly clouds.

When the pod door finally opens on Mars, the astronauts will not see quite such a jewel. Robots and automatic cameras have already been there, and the view is different. The first TV pictures relayed from a fly-by spacecraft showed nothing except a milky smear. The space scientists were momentarily dumbfounded. Was there nothing there? Only when they increased the contrast a hundredfold beyond normal did they make out any features. This characteristic martian haziness is due partly to dust in the atmosphere and partly to the uniform pale orange sand that covers all surface features. Allowing for all of this,

Mars is quite a planet. It comes with volcanoes, mountains, plains, craters, ice caps at the north and south poles, and one enormous canyon long enough to stretch from New York to San Francisco, and wide enough to swallow Massachusetts.

Mars's Mount Olympus is the twin to the caldera on Venus. It is 15 miles high, 350 miles wide at the base, and has a crater at the summit 35 miles in diameter. Right now the volcano is not active, but in the past it must have covered the entire surface with ashes and sent tremors to the four corners of the planet. The canyon is named "Mariner" after the U.S. space probe that discovered it in 1971. This gash is an enormous split in the surface caused by a slow local expansion of the planet. Close-up photographs show landslides which have occurred as cliffs pulled away and the rocky debris tumbled to the canyon floor. The craters with low circular walls are the scars of meteorite impact, some old, some more recent.

Lowell's long straight canals brimming with flowing water did not show up, nor were they expected. There is no liquid water on Mars, only a thin layer of ice crystals that settles on the rocks during the cold nights, and a deep permafrost underneath the sands. The air is so thin on Mars that even if a glass of water could be placed on the surface, it would quickly froth and boil away. The white polar caps are a thin layer of frozen carbon dioxide—dry ice. When the dry ice melts off in the Martian summer the temperature of the soil stays low, indicating that frozen water is mixed in the subsurface soil and rocks.

What about the dark areas, those olive-green "continents" mapped so carefully by telescope astronomers? Christian Huygens discovered the most famous one just after dinner at 7 p.m. on November 28, 1659. Upside-down through his refractor, it had the size and shape of the subcontinent of India. He named it Syrtis Major, Latin for "big bog," and over the centuries it was seen to expand and change as the "bog plants" grew. Close-up pictures show that these continents were nothing more than phantoms in the sand. On the beach at the seashore one can see dark and light colored particles, and these sometimes separate out by the action of the wind. The same thing happens on Mars. The sands of Mars are multicolored, and sometimes the dark particles are separated out by the wind and are blown over crater, mountain and plain to look to the distant earth-based ob-

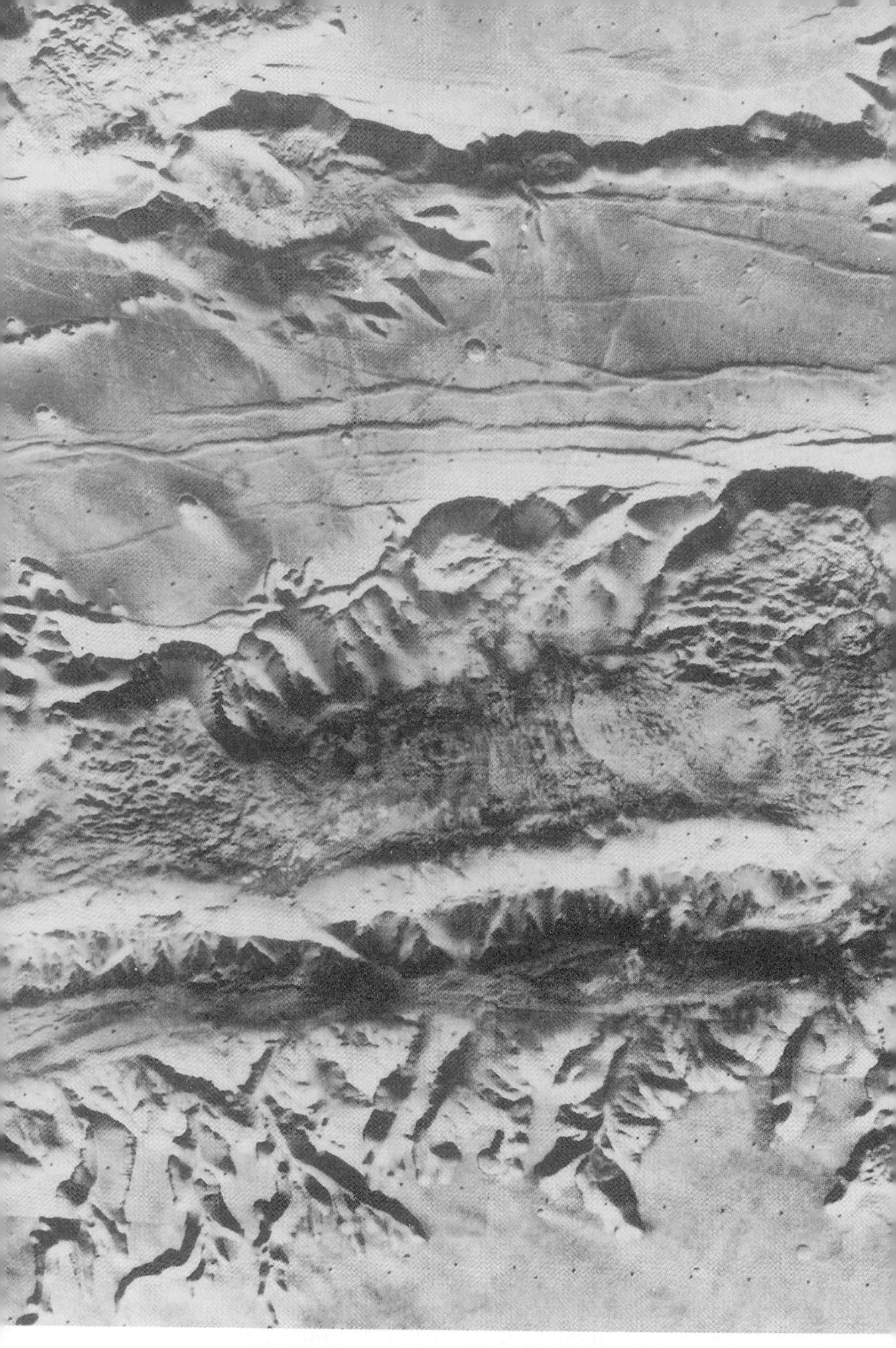

The Mariner canyon on Mars (north is at the top).

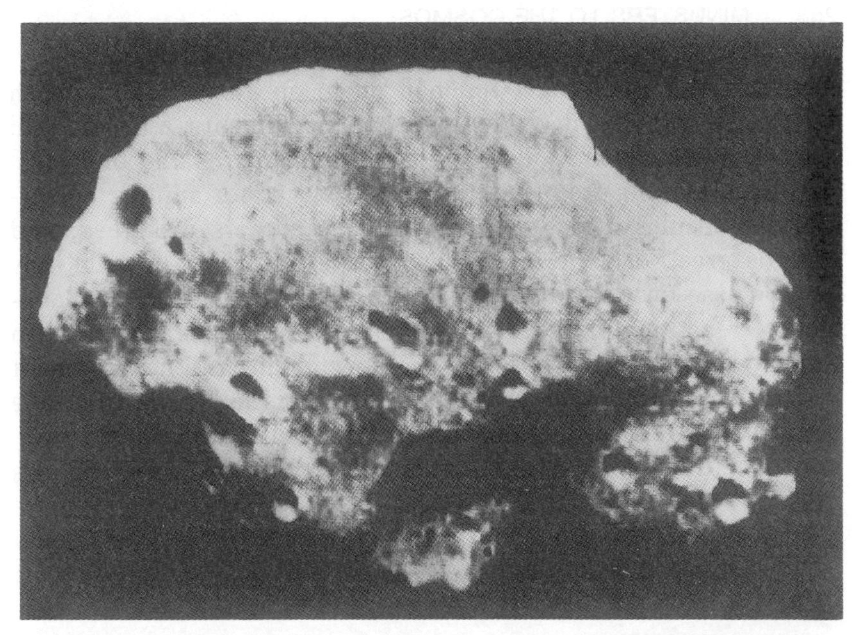

A close-up of Martian moon Phobos taken from space craft Mariner 9.

server like a great continent. The olive-green color was probably an illusion caused by refraction in the telescope, or color contrast. As Huygens might have said: "Sic transit Syrtis Major!"

What of the two moons of Mars, Phobos and Deimos (Fear and Panic), discovered with the U.S. Naval Observatory's 26-inch refractor in Washington, D.C., in 1877? A Russian science fiction writer claimed they were the hollow casings of vast artificial satellites launched by defunct Martians at the peak of their civilization. Close-up pictures proved the moons were no different than astronomers had imagined, rough irregular lumps of rock, pitted and marked by impact scars.

Life on Mars—was H. G. Wells right? No, there is nothing alive on Mars, not even a bacterium. Nothing is in the sand to show that life ever began on that arid next-door neighbor to earth, and no hope can be held out for life ever beginning spontaneously there in the future. A sweeping conclusion like this is not arrived at lightly. It was the majority opinion of the several hundred scientists who studied the data after Viking landed on the planet and carried out a biopsis with a robot spacecraft more capable and intelligent than any previously designed in the space program. When Viking I touched down on July 20, 1976, with retrorockets ablaze, it carried weather instruments, a mechanical arm for scooping the soil, a seismograph, cameras and biological detectors ranging from incubators to devices for sniffing out organic molecules. One or two of the researchers disagreed, and still disagree, with the conclusion. In their minority opinion they say the case for a dead Mars is not proven and that the explanation of reactions in the biological detectors as due to hydrogen peroxide were questionable. But the clincher in the argument which swayed the majority was the result of the organic molecule sniffer. No hydrocarbon molecules existed in the soil, dust or air, and the fundamental building blocks for life as we earthlings

Rocks, sand and a coating of ice on the surface of Mars as viewed from Viking 2.

know it were absent. Of course, life based on a different chemistry could be there, but nothing moved, no strange things were seen on the sand or rocks, and when the scoop dug into the soil there were no "worm tracks," no fossils nor other signs of life. Mars seems to be made of dead sand, rocks, dust and minerals. Not one live cell was detected, nor any organic molecules.

A ready explanation for this lifelessness is the lack of water and the sparsity of oxygen in the thin atmosphere. Mars had water in the geological past, but it is now locked up in the nighttime frost layers and in the deposits of buried permafrost. The maps obtained by orbital reconnaisance vehicles show dry riverbeds hundreds of miles long and a hundred feet deep. For a short

time in the past—a week or maybe a month—a sudden torrent poured along, sweeping debris before it, gouging channels and leaving islands behind. Then the water disappeared, evaporating into the air or finding a chasm in the ground in which it froze again. In one area the source of the water can be pinpointed. There is a field of jumbled rock about 100 miles across, and a large washway comes out from one edge. Probably there was a giant ice dome under the surface which was suddenly heated by volcanic activity. A subterranean glacier was unleashed and the water surged out, carrying all before it. The rocky crust collapsed into the hole, leaving behind a jumble of dry Mt. Ararats.

Long ago ice melted beneath the surface of Mars, and a deluge rushed out to the left.

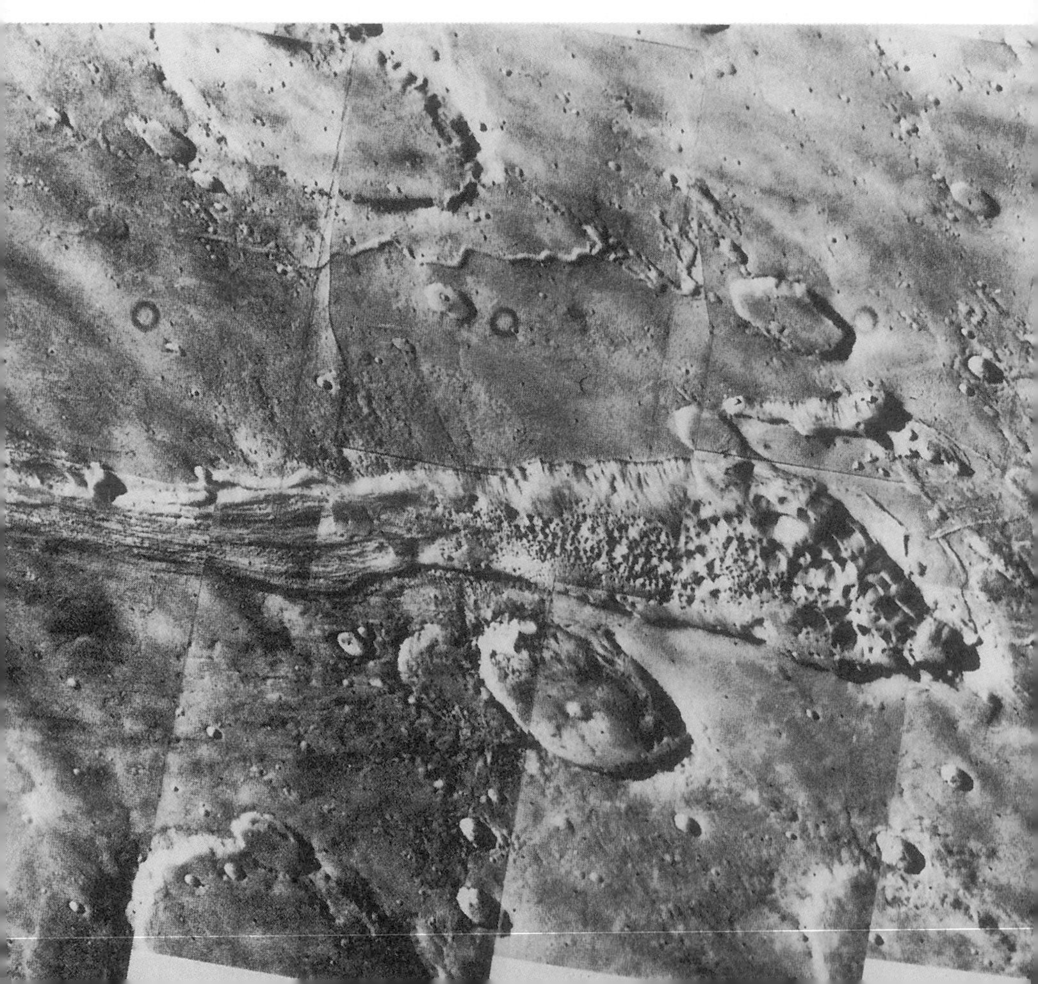

Water is a biological necessity for the beginnings and continued support of life. Mars may have had more water in the past than it has today, but the outpourings were brief, too short for the long period of growth and evolution that seems to be the pattern established on earth. A number of scientists think there is hope yet for Mars, and they pin these hopes on a future discovery. Carl Sagan says the Viking robots were smart, but not as smart as a grasshopper. By this he means the robot could not hop about. The 2 Vikings searched for life at only 2 spots on that vast planet. Life might exist elsewhere—at the edge of the melting pole caps for example. Then again, if Mars is dead now, there is hope for life in the future. Astronauts could set up greenhouse culture-pods, trapping the air and solar warmth and releasing the subterranean water. Mars could be seeded from earth.

Out from Mars, across the asteroid zone, probes have made their rendezvous with Jupiter. It was that planet which inspired Gustav Holst to write his symphonic Opus 32, *The Planets,* and it was that music which was played over the loudspeakers at the Jet Propulsion Laboratory in California as the first TV images came in from Jupiter, across the billion miles of void. It was a fusion of the creative arts and science. The astronomers' first reactions to the pictures were "No!" ... "Incredible!" .... They couldn't believe what they were seeing. Strange though it may seem, this is a common reaction of scientists. The results were totally unexpected, a shock, but the pictures were real, and the awed watchers knew that ultimately, somehow, they would have to overcome the feeling of disbelief and come to grips with the surprising truth.

The pictures showed a glowing, colored cloud system. The whole earth could be swallowed up in one of those swirling clouds on Jupiter. The Jovian atmosphere is thick and deep. It must be heated from below, because the planet radiates more heat than it receives from the sun. This heat comes from inside the planet, either from slow contraction or as heat left over from its formation.

The Great Red Spot turned out to be a persistent hurricane, the longest one on record. Seen by Huygens in 1659, it was going strong when Voyager flew by 320 years later. Thirty thousand miles across, the winds in this Jovian storm take about 6 days to rotate once around the eye of the vortex. Like a tropical storm on earth, it is fed by warmer air rising from the lower layers of

Satellites Io and Europa orbit above Jupiter's clouds as seen from Voyager 1.

A close-up of Jupiter's Great Red Spot, lower left.

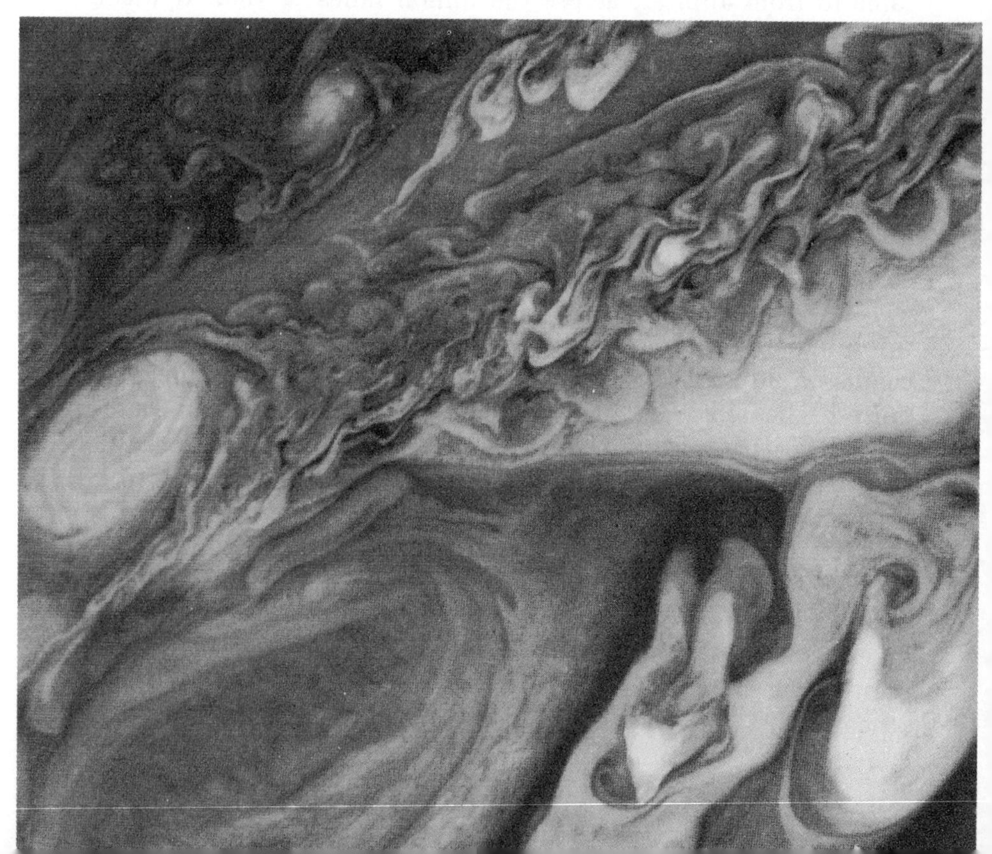

the atmosphere. On earth the warmth is solar energy stored in the oceans. On Jupiter the heat source is Jupiter itself.

Enormous thunderstorms rage on Jupiter. They can be seen most easily on the dark, nighttime side of the planet. The lightning bolts are thousands of miles long. The droplets in the clouds are ammonia and methane, colored by phosphorus and sulphur.

Jupiter has a faint ring system like Saturn's. The rings are made of chunks of solid material, rock and ice, moving in near-circular orbits, slowly spiraling into the equatorial regions of the planet. The rings are about 20 miles thick, but thousands of miles wide.

There is not much chance of finding life forms on Jupiter. Below the clouds it is as dark and cold as the earth's South Pole in winter, and above the clouds there is a radiation belt 100 times stronger than the lethal dose. These charged radiation particles interact in some yet to be determined way with the moon Io. As Jupiter rotates, a stream of electrified gases focuses on Io like the beam in a TV tube, and the current flow is 5 million amperes—enough to blow all the fuses in the city of Ames, Iowa. Deep down below the clouds Jupiter probably has a liquid or mushy surface of ices, equally lethal, with pressures thousands of times greater than those at the earth's surface.

Galileo was criticized for being an unimaginative scientist when he named the moons I, II, III, and IV. Others dipped into the classics to give them the names of Io, Europa, Ganymede and Callisto. Now, so many moons have been discovered by the space probes that scientists have gone back to Galileo's old system. At the latest count the moons go from I, II, III, and IV up to moon XV.

Telescope astronomers watched the 4 points of light move on each side of Jupiter like pearls on a thread. They assumed the moons were all alike. They are not.

Io should have been called Vulcan; it is the volcanic moon. As many as 8 volcanoes have been seen in violent eruption at any one time. And the volcanoes are big—craters 150 miles across, plumes of ejecta blasting upward into huge umbrella-shape clouds. What makes this little moon, Io, so volcanic? The action is due, apparently, to the shape of the orbit. It moves in a slightly elliptical orbit, and since it is so near to Jupiter huge tides are raised in the volume of the moon. These movements are like earthquakes, but induce from outside, not from within. The re-

Io, the volcanic moon.

sult of the tides is a bending and heating of the rock, followed by vulcanism. Another source of heat is the 5-million-ampere current from Jupiter.

Europa is the ice moon. Europa's surface is white and smooth with no meteorite craters or mountains. There are long cracks across the surface where water has been forced up in the past and has frozen again at the surface. Dark ridges of dirt run across the ice fields, and there are localized patches of dust and rock. NASA scientists believe the cracks could harbor microscopic ocean plants like those found beneath the ice in Antarctica. They say the conditions are similar, but so far there is no proof for Europa organisms.

Ganymede is the grooved moon. The surface of this larger moon is a dark terrain covered by long grooves as if it had been raked or harrowed. Early on in the history of Ganymede impact craters formed, but later the surface wrinkled, perhaps due to

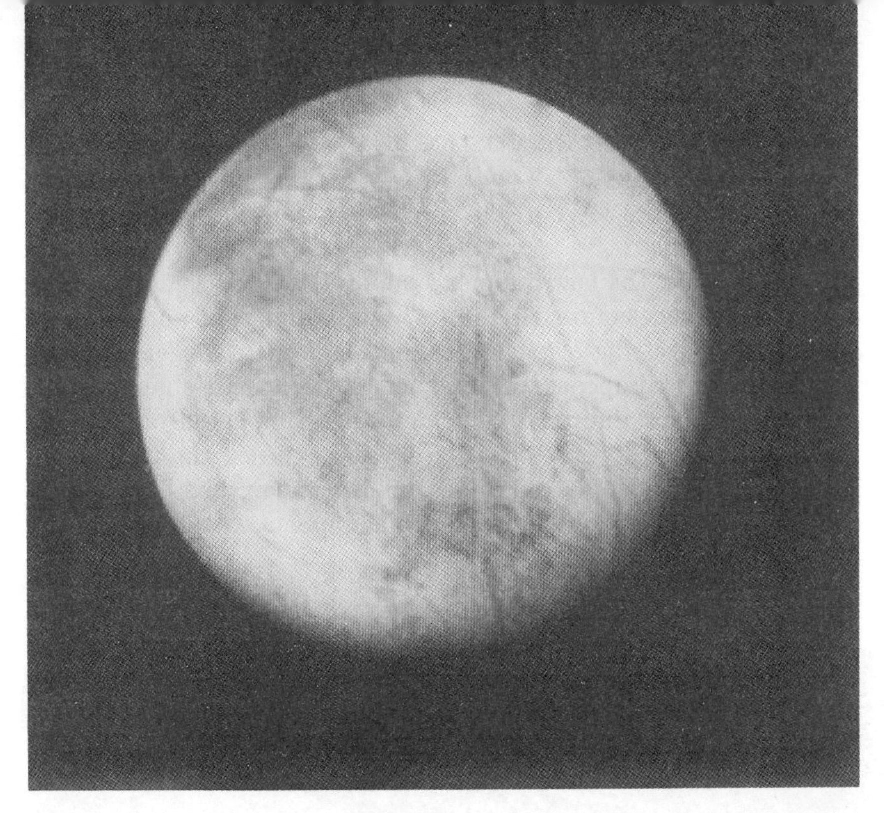

Europa, the ice moon.

A close-up of Ganymede, the grooved moon.

contraction of the icy core, to make vast, parallel ranges of low hills.

Callisto is the astronomer's moon—it fits the stereotype model expected before the space age, when astronomers were looking at the string of pearls. Callisto is a little bigger than the earth's moon, composed of solid ice, spherical in shape with a covering of dark debris. When meteorites impacted on Callisto they blasted holes in the outer cover and exposed the whitish ice beneath. It looks something like our own moon, but because the ice melted and froze again after the impacts, the floors of the

Callisto, the astronomer's moon.

The first approach to Saturn.

craters are smooth. If ever there are going to be Callistilians, they will have superb skating rinks.

Saturn, the planet on the edge of the classical world, is very much like Jupiter. It is cooler and smaller and farther from the sun, but it has thick cloud belts, moons and rings. As with Jupiter the classical naming of the moons was abandoned. After Titan, Tethys, Enceladus, Dione, Mimas and Iapetus, new discoveries were given a plain S-number (S for Saturn), and the last one added was S-16. Two of these moons are in the same orbit and it is a wonder that they don't collide.

Titan is the hydrocarbon moon. An orange smog hung over the surface each time a close-up picture was taken. Its atmosphere is mainly nitrogen with 1 percent methane, and the smog is made by aerosols colored by the deadly chemical hydrogen cyanide. These chemicals come from the primordial methane. A polymer chemistry has been going on in the atmosphere for 4 billion years, and a deposit of ethylene, ethane and other hydrocarbons now covers the surface to a depth of a mile or more. These chemicals will burn in an oxygen atmosphere. There is enough fossil fuel on Titan to light up the earth for centuries. The only difficulty is getting there and bringing it back.

Mimas is the mini-moon with the maxi-crater. One large impact crater stands out on the surface looking like the top of a goldfish bowl. How such a small moon could survive such a large impact without shattering no one knows.

Iapetus is the Yin and Yang moon—the forward side is dark and the trailing side is light. The black and white coloring probably has nothing to do with the motion of the moon in its orbit, but is caused by an unusual separation of its rocky and icy parts.

What about Saturn's rings? Galileo with a magnification of 32 could only make out what he thought were appendages. Even with the most powerful telescopes the rings looked completely smooth, like continuous halos surrounding the planet. French astronomer Giovanni Cassini discovered the dark division in the rings which separated the halo into two rings, A and B. But a thousand rings were seen on the space-probe pictures. The effect was like looking at the grooves in a phonograph record. Nor were the rings exactly circular. Some were oval, and some twisted, and two appeared to be braided together.

Of course, the rings cannot be solid; they would break up under the gravitational strain. They must be a whirling cloud of particles, but close though the probes were, the cameras could not resolve the individual objects. Earth-based radar reflections show these myriads of moonlets to be bits of ice and rock ranging in size from a few inches to a few feet.

There must be forces other than the force of gravity at work in these thousand rings, otherwise the rings would be the smooth discs expected by astronomers. Saturn has a strong magnetic field and a magnetosphere. It is possible that a complicated gravitational, magnetic electrostatic interaction is molding the shape of these rings.

Uranus, Neptune and Pluto have not yet been visited by space probes. One of the Voyagers, having passed Saturn, is due to skim Uranus on its way out of the solar system, and if the TV

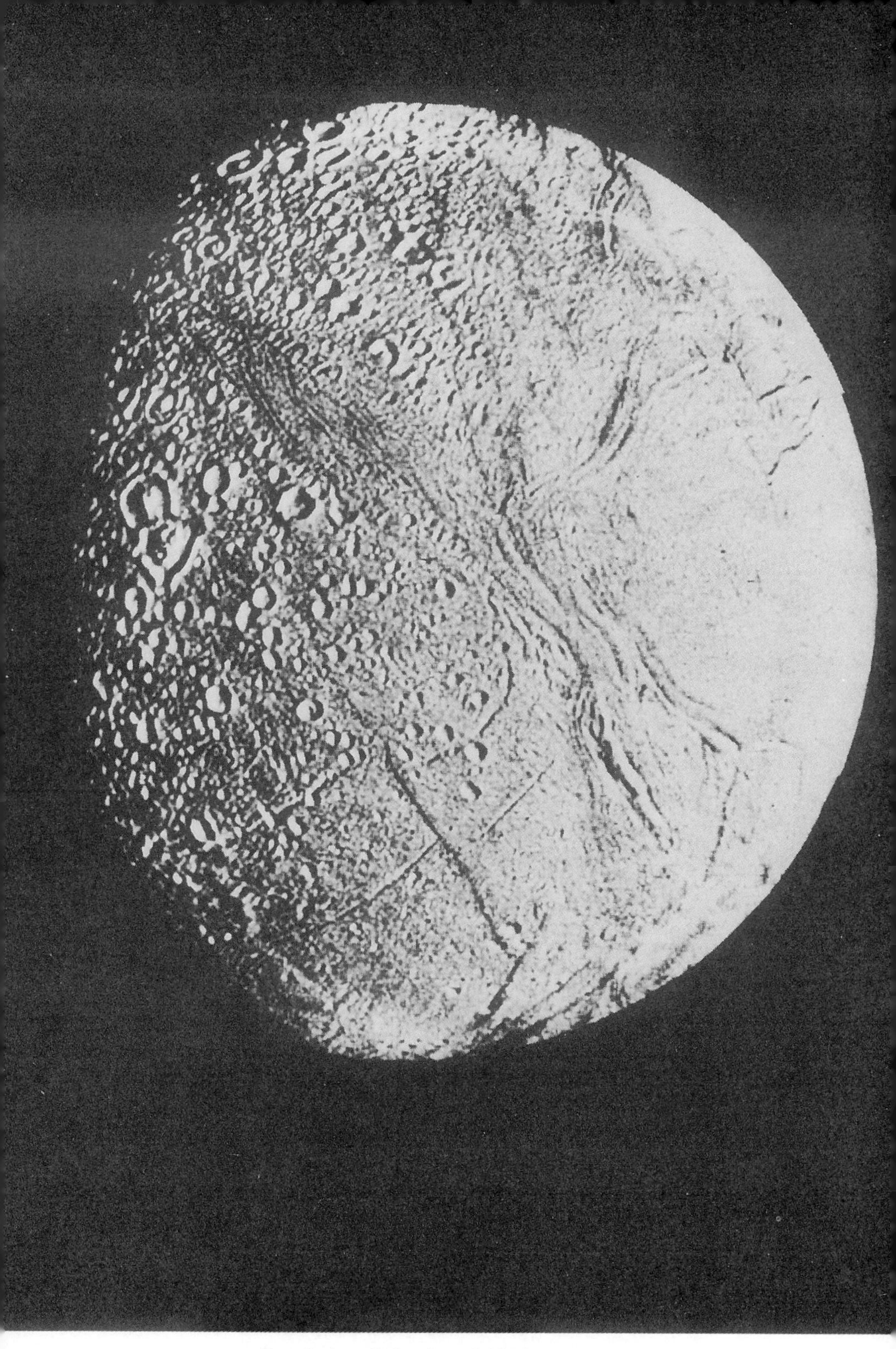

Enceladus, Saturn's wrinkled moon.

A close-up of Saturn's rings.

cameras are A-OK at that time, then we can expect to learn a lot. In 1977 a conventional astronomer made what might prove to be the last important discovery with earth-based equipment. He measured the light from a distant star as Uranus passed in front. Result: the light faded in a rhythmic pattern. Conclusion: Uranus has a ring system made of dark, nonreflecting objects which momentarily had cut out the light from the star. Perhaps the Voyager spacecraft will tell us more.

A grasshopper is more intelligent than the Viking lander, but a man is more intelligent than a grasshopper. NASA has always argued strongly for a manned space program. The Soviet Union plans for manned exploration of the solar system. With the shock of Sputnik and other Russian successes (the Soviets had already orbited a live animal before the first U.S. satellite was off the ground), President John F. Kennedy charged NASA with the task of placing an American on the moon before 1970. With splendid timing, on July 20, 1969, Neil Armstrong landed the "Eagle" on Mare Serenitatis. "One small step for man, a giant leap for mankind," he said as his corrugated sole made a footprint in the loose dust. More correctly he should have recognized our species in general terms—one small step for earthlings—but the world as it listened knew well what he meant.

Personally, I was nervous as I watched. Thomas Gold, astrophysicist at Cornell University, had predicted a possible swallow-up. The dust of Mare Serenitatis could be half a mile thick, he said, soft and fluffy like dry cement powder with no bearing strength. If so, the astronaut would sink out of sight, leaving only a smooth, tubular hole. It was only a theory, and fortunately it was wrong. The dust layer was a thin sprinkling over hard, compacted rocks. Armstrong's footprint sank in about half an inch, and the dust spread out from the impression smoothly like the cement powder that Gold had predicted, but the rocks underneath could hold the weight. If future space travelers do not disturb it, that footprint will stay fresh for millions of years until it is covered over by new dust from meteorite impacts.

The moon's surface turned out to be ideal for getting around. Several later teams walked about on long safaris. By the time of the last expedition (Apollo 17, December 1972), an astronaut and a geologist were able to drive around in a pneumatic-tired moon buggy that would have done credit to Henry Ford. They breathed inside air-conditioned space suits, and lived and

"One small step for man, a giant leap for mankind."—Neil Armstrong on the Sea of Serenity

The mountains at the center of crater Copernicus.

worked on the moon for 22 hours, 5 minutes and 3 seconds of earth-time.

What was learned from this? First of all, there is no water on the moon, no liquid, no ice, no vapor. Earth rocks contain 2 percent of embedded water of crystallization (volcanic out-steaming was the source of the oceans in the past), but moon rocks do not contain even a trace of water. They are hard-baked, tectonic material, un-earth-like. Low-boiling-point metals such as sodium and potassium are scarce in moon rocks, again showing evidence of a hot past.

The date of solidification of the various moon rocks can be found in the ratio of radioactive decay products that are frozen within them. Rocks from the mountainous regions are older than the lavas of the plains. Some of the oldest rocks solidified at the time the planets of the solar system condensed, about 4.6 billion years ago. The lava, on the other hand, is about 3.8 billion years old.

These ages fit in with what must have been happening at the time the planets condensed from the embryonic solar nebula. In the very beginning, the meteorite chunks left over from the condensation crashed into the moon, making the overlapping craters of the highland mountains. Later a few large asteroid-size objects blasted out the large Maria which filled with lava to make the smooth plains. This was in the first billion years. Then the meteorites were mostly used up. Meteorites are still falling on the moon today as they are on the earth, but less frequently. These more recent craters are the source of the dust. Nothing happens on the moon except for these rare impacts when material is ejected to cover the surrounding surface. Occasionally large chunks of rock are thrown out. An Apollo team explored one of these chunks. They found it was over 20 feet high.

All this goes to prove that the moon was never a part of the earth. One theory had the moon breaking off from the equatorial regions of the earth at the time when it was spinning fast. But the moon rocks are not like the earth's crust and this theory can't be true. Something is very different about the moon. It has had an intense heat treatment sometime in its history. The moon may be an interloper; it may not be "our" moon at all! In many ways it is like sun-baked Mercury. Could it have formed close to the sun and then by gravitational interactions have moved out

Astronaut Harrison Schmidtt examines a moon boulder.

and been captured by the earth? Computer theorizers are trying to figure out a way for this to have happened, but so far a mathematical solution to the problem has not been found, and the origin of the moon is still a mystery.

The moon does not have volcanoes like Jupiter's Io, but it does have moonquakes. On their working visits the astronauts set out seismographs in various spots, and these instruments sent back continuous data. Moonquakes are much smaller than earthquakes—they would be unnoticeable even for a person standing right over the epicenter—but they are far more numerous, averaging for that small globe close to 3,000 per year. The moon's orbit is elliptical, and tidal forces from the earth raise and lower the rocky layers, causing them to bend and crack. The same thing happens on Io, except the tidal forces of Jupiter are more severe.

Like the other planets and moons in the solar system, the moon is dead. It has no air or water. The only oxygen is the little puff which out-gassed from Alan Shepard's golf ball when he took out his club and drove a hole (crater?)-in-one. All the precautions to quarantine astronauts for 10 days upon return from a lunar mission were unnecessary—there are no bugs on the moon. Even if bacteria did accidentally escape from the Apollo space suits (and every precaution was taken to ensure this did not happen), those bacteria would quickly have been killed by the intense ultraviolet light of the sun. If any life ever develops on the moon, it will have to be placed there and nurtured by life from mother earth.

THIRTEEN
Mindstep 4—Far from Home

It's a long way from earth to infinity. The more we know, the less we understand. Facts about fundamental particles and the true nature of gravity elude the human mind. Our earth as a home shrinks with each new discovery of the Space Age. Alvin Toffler's book *Future Shock* warned us that the most dangerous attitude of humanity was to ignore the changes swirling about us and to complacently dwell in the past.

There is no doubt that, like it or not, technology is here to stay. These new technologies extend the reach of the mind to the ends of space and the beginnings of time. Robot space vehicles have now been to the moon and all of the 5 planets known to antiquity. What was discovered was entirely unexpected—a series of shocks, not the least being the total lack of life, neither spores nor organically produced molecules. The extra-terrestrial solar system is an awe-inspiring, inert, inorganic desert.

Plato and Aristotle in antiquity had already talked of two worlds, the seen and the unseen. Aristotle believed there was an all-pervasive, unseen, creative intelligence. Before Ptolemy's time, the planets in the zodiac were the wandering spirits, but with one stroke he made them things. Ptolemy pushed back the realm of the spirits to the edge of the celestial sphere. The planets in the space between earth and heaven were real objects, inert and tamed by ordered motions. If a spirit world existed in the

Ptolemaic system, it had to be placed beyond the planet Saturn, at the zone of the fixed stars.

Copernicus continued the attack. He showed the stars to be other suns, and so the idea of the celestial sphere disappeared. Giordano Bruno said the field of stars was 3-dimensional and stretched to infinity, and beyond infinity was another infinity. (He probably would have agreed with my theory of the eternal universe.) Galileo's telescope, made of good Venetian glass, took his mind's eye far from earth, but everywhere he looked he saw only physical, earth-like objects. He could not see the spiritual world out there, and so he argued it must exist only in the mind. But for others what was seen in the telescope was only one part of the cosmos, the other part of the cosmos was out there, but unseen.

The bicameral brain of humans divides the cosmos into that which is seen and that which is unseen, the factual and the intuitive. Philosophers Immanuel Kant, Rene Descartes and Gottfried Leibnitz said duality was a fundamental to the universe itself. Scientists tend to look at one cosmos, humanists at another, and the two can never be joined. In the days of Gilgamesh, in mindstep 1, it was a split between the mortal and the immortal. In ancient China the philosophers recognized the inert matter of Yin and the creative force of Yang—the wooden log and the lively flame. Today we recognize the pod of the solar system and the life force, the earthling seed, within it.

At spring tide the pounding surf of the North Sea can be heard at St. John's Church, Great Yarmouth. Inside the church, over the nave, is a Gospel message in golden letters that I read as a boy each week during the sermon. It said that God was a Spirit and must be worshiped in Spirit and in Truth. I now know there are several deep interpretations of St. John's theology, but at the time those words seemed to me to point to provable truths versus something that went beyond. Photographs of stars and galaxies gave only one-half of the story, the material half.

I do not know who was the first astronomer to dream of extra-terrestrial (E.T.) life, but the thought is constantly there. Sir William Herschel believed there were people living on the cool surface of the sun beneath the incandescent clouds of the photosphere. Johannes Schröter thought he saw signs of industrial activity on the moon, and Percival Lowell dedicated his life

Jodrell Bank's 250-foot radio telescope.

to the elusive Martians and their canals. Yet an E.T. is a belief, not a fact. E.T.s exist only as an idea in the right hemisphere of the brain. Until now the search for proof has gone unrewarded.

That is why 20th-century space exploration has given pause. The U.S. Viking team, of which Carl Sagan was a member, hoped, almost expected, to find life on Mars. In previous years, Apollo scientists thought that bacteria might be brought back from the moon by the astronauts, so they all had to go into quarantine for 10 days when they got back home. The idea was that moon bugs might be dangerous to earthlings. But there was no life on the moon or Mars. No bugs, nothing. At the latest count, all the planets and their satellites are without life. Earthlings are alone!

What about other civilizations on other planets going around other stars? There are billions of stars and billions of galaxies; surely there are other civilizations out there! That line of argument is the rationale for the international scientific project called SETI (Search for Extra-Terrestrial Intelligence). The atmosphere is transparent to radio waves, and so a radio astronomer with his equipment sees out from earth as if he was already living in space. With modern technology we can begin to break the barriers of space communication. Radio signals give a direct connection. Calculations show that earth-made TV and radar signals could be detected at about 100 light years' distance with conventional low-noise receivers, and so earth should also be able to detect other civilizations. The theory is that "they," the "E.T.s," if they exist, have developed the same stage of radio technology that we have. They may even have surpassed us. After all, it was only 1901 when Guglielmo Marconi sent the first faint dots and dashes across the Atlantic from Land's End in Cornwall to Newfoundland.

In project SETI, radio astronomers have listened for intelligence on the hydrogen line frequency and other frequencies with the biggest dishes available. They have turned their antennas in the direction of more than 50 stars, all chosen as likely candidates. They picked out G-stars like the sun, single stars—not members of a binary system where the second star would interfere with the suspected life on the suspected planet—and all less than 100 light years away from the earth. So far the result has been negative. With all the long SETI listening, the only message received from those remote depths of space has been the steady, natural hiss, the random noise of hot electrons.

Nor do we have any positive proof that other solar systems exist. Admittedly the astronomer's earth-bound telescope is limited in range, and one can argue that surely with all the billions of stars in the universe, some at least must have planets, and some of these must be like the earth. But so far in their search astronomers do not have any proof for the existence of any planets going around other stars. Not one planet has been detected beyond our solar system. There were reports some years back of a star moving in a wavy line. The disturbance was supposed to be caused by a Jupiter-size planet. Later and more accurate measurements showed the results were wrong, or at the very best doubtful, and therefore the detection of the planet was "unproved."

Part of the problem in detecting other solar systems is observational and can be overcome. The stars are so far away, the

(Above) The U.S. National Radio Observatory's radio telescope, Green Bank, West Virginia. (Opposite) The U.S. Space Shuttle.

image is faint and small, and the earth's atmosphere disturbs the measurements. Maybe an extra-solar planet will show up in the future with the use of advanced instrumentation carried in orbiting observatories, or from a scientific base camp set up on the moon. Meanwhile the hard fact at this time is that there are no detected signals and no detected planets.

What about other galaxies? With the present state of the art, we have the capability of detecting what can be called super-technologies, even if these are as far away as the Andromeda

USA

galaxy. If a group of intelligent beings in a super-technology wanted to send a message across the 2 million light years between us, and if they were willing to use a good fraction of their available energy resources in a burst of radio waves, then no technical reasons would stand in the way. We on earth are already equipped to receive those signals. The first search of Andromeda gave a negative result—no signals other than the thermal electron hiss. And in this search the radio astronomers took in at one glance in their antenna beam a system of 100 billion stars—none, apparently, with a super-technology present.

Negative results or not, the SETI search will go on in the U.S. and around the world. Other frequencies, other possible modes of E.T. signaling, will be monitored. For example, E.T.s might use laser light, a distinctive and powerful invention which sends photons in a narrow, coherent beam, or they might use X rays or even neutrino pulses. It is not impossible for an advanced E.T. civilization (if it exists) to send a probe vehicle to circle in orbit about the sun, collecting data and signaling its presence.

Back in 1967 there was what seemed to be a detected signal, and for a few brief months the antennas of radio astronomers around the world turned on one small spot in the sky like compass needles. An Irish-born graduate student at Cambridge University in England, Jocelyn Bell, had just finished 2 years of swinging a sledgehammer, driving stakes into the ground to make an array of 2,040 antennas. (In those days of slim budgets, the rule in England was "make your own antenna, then use it.") When she used it she found radio clicks like the sound of a grandfather clock coming from one part of the sky. The pulses were regular, very regular. The only change was the slight shift caused by the swing of the earth in its orbit around the sun, proving the pulses came from beyond the solar system. Could it be the LGM—the little green men—she wondered. Radio astronomers swung into action. The stream of discoveries was shared by letter and telephone, because the publishing scientific journals could not keep pace with the events. At the height of the investigation the large Arecibo antenna, built in a limestone sinkhole in Puerto Rico, showed a jumble of information carried on each pulse, like the lines of a TV program: LGM-TV.

What Jocelyn Bell had discovered was a new class of object—pulsars. Within a year more than a dozen had been found, and now the total is close to 200. A pulsar is a star one stage more compressed than a white dwarf. It is a neutron star. In a white dwarf the electrons touch. In a neutron star the electrons join with protons to make neutrons, and the neutrons touch. These neutrons are squeezed together so tightly that the star is no more than 12 miles in diameter. Because of the peculiar physics of neutron material, the outer shell is a crystalline solid and the inner core is fluid. Having shrunk down so much, the star is now spinning very rapidly indeed. Jets of ionized material spin around with the star, and the pulses of radio energy come from these jets. In the case of pulsar No. 1 the pulses sounded like the ticking of a grandfather clock, coming precisely every 1.3 seconds. Other pulsars are quicker, equaling the ticking of a wristwatch. One pulsar is flashing on and off at the rate of 30 times per second—so fast that it would appear like an ordinary unblinking star to the naked eye. Special photo-optical devices are needed to monitor the pulses. This fast pulsar is at the center of the Crab nebula. Although they did not know it, what those Chinese astrologers recorded in 1054 A.D. was the birth of a neutron star, as the total content of a white dwarf collapsed into neutron degeneracy. This spinning beacon in the Crab nebula has caused the gases to shine ever since.

Pulsars are the most regular thing in the stellar population. Their spin is more reliable than the spin of the earth or the tick of a clock. The Crab nebula pulsar loses only 15-millionths of a second in a year. This slowing down is caused by the bleeding off of energy within the swirling ionized jet. Viewed from earth a pulsar is like the flashing light on a tow truck, or, more poetically, like a rotating lighthouse beacon. Occasionally there is a small, sudden change in the period of a pulsar, and then it settles down again to a regular spin. When this happens the crystal lattice has cracked and readjusted. The pulsar has suffered a starquake.

Some neutron stars are member of a binary system. When a neutron star and a normal star make a spinning double, gases are blown off the shell of the normal star and move in the intense gravity toward the neutron star. The gases ionize and funnel down magnetic field lines to make intense hot spots. The neutron star emits X rays at these spots, and as the star rotates, X-ray pulses are received on earth. But for the high-altitude observatories flown above the absorbing atmosphere in satellite spacecraft, these objects would go unnoticed. Now they are watched and studied as a clue to the final ending in the life of a star.

New technology produces new surprises. "I went home in a state of disbelief," said Palomar-based Maarten Schmidt. It was 1963, and the Dutch-born astronomer had just measured the redshift for the first known quasar. In the previous year, Britisher Cyril Hazard had pinpointed radio star 3C 273 by watching it fade as the moon passed in front. What astounded Schmidt was the distance of 3C 273. If the redshift was caused by the expansion of the universe, this object was 2 billion light years away. To measure it at all on his photgraphic plates, it had to be one of the brightest, if not *the* brightest of objects in the whole cosmos.

What could it be? Radio star 3C 273 was not alone. Very soon other quasars were tracked down in the multitude of specks in the telescope photographs. Hundreds of them were detected and measured, with always the same result—super-luminous objects on the far frontiers of the universe. The latest record-breakers are at distances of 15 billion light years, and they are brighter than 10,000 galaxies like our own put together. And they sparkle; the brightness changes from week to week. This means the source of this intense light must be small, no bigger than our solar system. Is it possible that astronomers are being fooled by what they see? Could something else be shifting the hydrogen lines from the blue to the red end of the spectrum? Might the quasars be close by and therefore not so astoundingly bright? A strong gravitational field can displace lines to the red, but no one has been able to prove that this is the mechanism for the quasars.

No, the general conclusion is the almost impossible: we are looking to the far edge of the known and knowable universe, to the very beginning of time when the big-bang fireball was in its earliest stage. There are no quasars nearby. The galaxies in our local group seem to be the same age as the Milky Way, close to 20 billion years old. If quasars did exist in the local group when the Milky Way was formed, they must have burned out a long time ago. Quasars seem to be enormously bright, short-lived phenomena. They are something that existed for a brief flash at the time of creation.

With high-powered telescopes, quasars sometimes show faint arms and halos. If these appendages are made up of stars as with normal galaxies, then the quasar could be a super-bright nucleus. Maybe quasars are galaxies that went wrong, malformed at creation. There is evidence for explosive ejection of material from them. Some quasars look as though they are blowing up. It is difficult to imagine planets or extra-terrestrial intelligence existing under such violent conditions.

When we look at a quasar, we see something as it was before the earth was formed. When we look at Andromeda, we see a galaxy as it was before Homo sapiens walked the earth. The boundary of the universe has been pushed to the limit of the range of the photon, and to begin to understand what is out there the mind must also be pushed to its limits.

One of the strangest forms of cosmic matter existed for 50 years only in the mind. The German mathematician Karl Schwartzchild took Einstein's new theory of relativity in 1916 and played a mathematical game. If a star, or a planet or a galaxy could be compressed to a pinhead, it would make a black hole. The equation was straightforward, but it was a big "if," and the professor doubted that this mathematical monster could exist in the real world, beyond the confines of his mind.

Einstein's equations predicted, in prose, for a black hole that:

> The material shrinks down to a mathematical point,
> smaller than a pinhead.
>
> Gravity is so strong that rays of light go into orbit.
> Photons whiz around forever and can't escape
> in a sphere that is called the "mass horizon."
>
> Outside the black hole space and time are normal,
> but inside they are not.
>
> A glitch has been created in the smooth universe,
> something far worse than a vacuum.
> No light, no energy, nothing can ever leave it.
>
> The black hole becomes a cosmic Pac-Man,
> swallowing up dust, gas, planets, stars and galaxies
> without leaving a trace.

Lewis Carroll's mind gave birth to something very similar in 1876. He called it a "Boojum." According to Carroll it was all right to hunt the mythical beast called the "Snark," and he described how a sailing vessel set off to catch one. The only risk was if the Snark turned out to be a Boojum, because those things had the odd knack of making people disappear when they got too close. As his verse explains:

> Erect and sublime, for one moment of time,
> In the next, that wild figure they saw
> (As if stung by a spasm) plunge into a chasm,
> While they waited and listened in awe.
>
> "It's a Snark!" was the sound that first came to their ears
> And seemed almost too good to be true.
> Then followed a torrent of laughter and cheers,
> Then the ominous words "It's a Boo—."
>
> In the midst of the word he was trying to say,
> In the midst of his laughter and glee,
> He had softly and suddenly vanished away—
> For the Snark was a Boojum, you see . . . *

The size of the black hole–Boojum depends on its mass. For a point with the mass of the earth, the horizon is only an inch across. If the earth could be squeezed down to the size of a golf ball it would never spring back. Nothing could stop the instant collapse to a pinhead surrounded by the never-never space of the black hole. A star collapse makes a hole about 10 miles across. A galaxy collapse makes a black hole slightly larger than the solar system.

Nothing can break up a black hole, so the equations say. Once it forms, it exists permanently in space. Nothing can escape from a black hole. Gas, planets and stars approaching it whiz around in a frenzied orbit, finally dropping into the black hole. Ordinary time stops at the boundary. Although the final collapse takes no more than a few seconds, gravity destroys the relationship of space and time. An observer looking at the event from the outside sees everything in slow motion. Time freezes at the instant the black hole forms. Light vibrations slow down to zero. Eternal blackness sets in.

For 50 years Schwartzchild's black hole was nothing more

* *The Complete Works of Lewis Carroll* (New York: Random House, 1976).

A radio map of the center of the Milky Way galaxy.

than a mathematical exercise. It was a Boojum of the left hemisphere of the brain. But then this phantom of the mind had to be faced in the real cosmos. Evidence came along in the 1960s for the existence of black holes near other stars, in globular clusters and at the center of our own Milky Way. Creative Yang had found a way to go beyond the compression of a white dwarf, beyond a neutron star. If the physical conditions are right, the temperatures and crushing gravity of a dying neutron star can trigger the final collapse into Yin-oblivion.

If the hole is black, how can we see it? The objects give themselves away by X rays. Material whizzing across the boundary into the black hole sends out a burst of X rays. There is an X-ray source in the constellation Cygnus, designated Cygnus X-1. The object seems to be a star with a black hole as a companion. The normal star is being swallowed up layer by layer. Measurements of the orbital period show the black hole is about 6 times as massive as the sun.

Other X-ray sources show up in the center of globular clusters. Astronomers think these black holes are larger, equal to the mass of a thousand stars and filling a volume as big as one of the moons of Mars. Smaller black holes might have bumped together at the center of these tightly packed clusters to make the large one.

Our own galaxy has a bright X-ray source at the center. Astronomers think it might be a black hole. It can't be seen in the telescopes because the dust and gas in the spiral arms block the view, but the radio pictures show a point with 3 small whirling arms. Will this black hole destroy the Milky Way? Not so long as the galaxy rotates and the stars in the spiral arms keep a safe distance.

The more we know, the less we seem to understand the cosmos, and our relation to it becomes insignificant. In the last 10 years a few astrophysicists have rebelled against this shrinkage, this loss of status. They have spearheaded anthropic principles I, II and III. . . .

I: The universe could not exist if intelligent beings were not here to observe it. This principle is more like a belief than a provable theory. The idea draws on fundamental philosophies such as those of Plato and Descartes. Nothing exists, says the principle, if it cannot be linked with a cognizant mind. To paraphrase Descartes: "The cosmos is seen, therefore the cosmos exists."

II: The physical laws of the cosmos are arranged to produce intelligent life. Gravity, quantum mechanics, chemical bonding rules, are there for the purposes of biochemistry. It is no accident that gravity triggers nuclear energy within the stars but leaves the smaller planets cool for life. Gravity is a weak force, designed to enable a planet to circle its star at a safe distance. Electrical fields are strong in order to bind the atoms in a molecule together. If these forces were the other way round, the earth would drop into the sun, and the DNA molecule would fall apart. Because of the laws of nuclear physics, the formation of the elements in the expanding big-bang fireball stopped short at carbon. How could the heavy elements so necessary for life be produced after that stage? They are produced by the accidents of the laws which lead to the red-giant phase, the helium flash and the subsequent collapse to a white dwarf. Then the carbon atoms are spewed out into space by nova and supernova explosions to provide material for the planets. But this is no "accident," says the highly controversial anthropic principle II; it is a built-in cosmic design.

There are those who argue for, and those who argue against. The choice falls between a cosmos of atoms and objects where intelligent life is unconnected with the stars, and a cosmos where intelligence has a place. Many astrophysicists see the stars as exploding, pulsing spheres of gas with no relation to life on earth. But others ask now if there is a cosmic connection. Are we as earthlings uninvited observers, or are we participants? Here the line between proof and belief, science and religion, seems to have softly and suddenly vanished. Freeman Dyson of the Institute for Advanced Study at Princeton University said at the centennial celebration of Einstein's birth: "All this sounds to a contemporary astrophysicist vague and mystical. But we should have learned by now that ideas that appear at first sight to be vague and mystical sometimes turn out to be true."

III: The third anthropic principle says quite directly, "Because we are here, the cosmos is altered." The observer, be it earthling, andromedoid or galactan, has an effect on things and influences developments. At the smallest level a physicist cannot make measurements of an electron without disturbing its track. At the larger level, intelligent beings throughout the cosmos can use the laws of nature to change the cosmos.

In mindstep 4 the earthling pod door opened. Humans have walked the surface of the moon, and humanoid robots have reached all the planets known to antiquity. Spacecraft have left the gravity of the solar system and are now speeding with earthling messages to the nearest stars.

No one knows where mindstep 4 will lead us. Biologists are quick to point out the basic fundamentals of all living organisms. Reproduction—preservation and transmission of the genetic code—is one fundamental, colonization is another. It is a rule of all species known on earth to expand to the limits of the environment. All life forms, from bacteria to butterflies, from protozoa to primates, from hydra to humans, colonize. There are ant colonies, bird colonies and beaver colonies. Will Homo sapiens colonize space? Some people think so. Homo sapiens is the only species equipped to do so, and space is there. NASA has funded a study which is now in progress. The U.S.S.R. set out on a goal of a manned orbiting space station right at the start with the Sputnik tests of 1957.

Tsiolkovsky was one of the first to talk about space colonies. He dreamed of giant floating greenhouses which contained water and soil for food production. This was before the communist revolution, before radio, telecommunications and computers. In his mind he saw space adventurers leaving earth to capture a small asteroid. The stone, iron and minerals in the 10-mile-wide object

The cryptic earthling message dispatched on Pioneer 10.

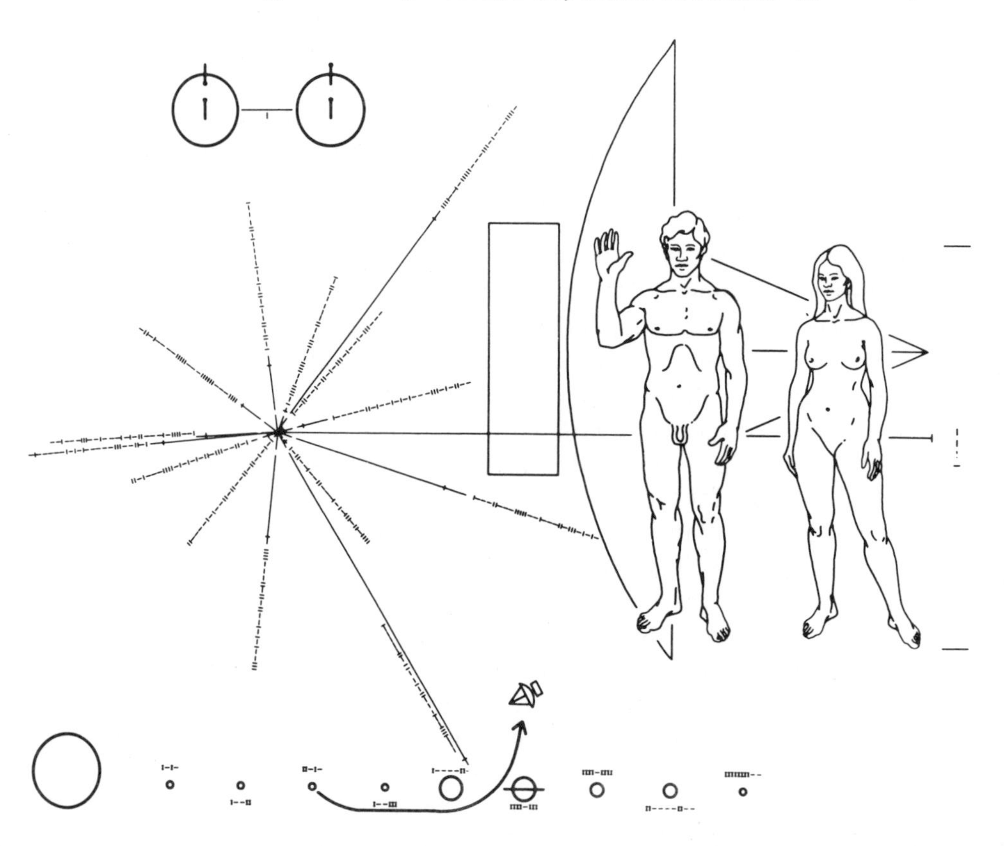

Pioneer 10 escapes from the solar system.

would be mined and processed to make the greenhouse and the soil. Those early ideas colored the whole Soviet space program.

In the United States the talk is of mining the moon instead of an asteroid, using solar energy to extract oxygen from the rocks and to smelt out metals. If by the year 2050 A.D. the earth is overcrowded, fuel-depleted and atmospherically polluted, then a self-contained capsule for 10,000 or 20,000 souls would be a haven of escape. All the manufacturing processes would be more simple in space, so the engineers say, because there is no gravity, and the supply of solar energy is unlimited.

Some engineers do not look on space colonies as a haven, but as a humanistic challenge. Like in the good ship *Enterprise* of "Star Trek", there is for them the quest of exploration. Bases would be set up on the moon and then on Mars. These would be way-stations for the long journey to the stars. Saturn's moon Titan has already been earmarked as a filling station for free hydrocarbons, and nuclear fuel can be added to the propulsive mechanisms for deep space flights. Hopping from one environment to the next, Homo sapiens could colonize the entire Milky Way galaxy in 100 million years, according to the planners. That, they say, is human destiny. From the earth-pod the cosmos will be conquered.

But, of course, theories are theories, and no one can predict the ultimate limit of mindstep 4. Ptolemy could not foresee Copernicus. Copernicus could not foresee a return to an earth-centered universe with hundreds of satellites in geocentric orbits. Earthlings have problems which seem impossible to solve. Not the least is the aggressive gene in our species. If nations cannot cooperate on a small globe of rock and water, could they co-exist in space? Who would control a space colony? Would there be a federation, would there be peace, or would star wars be the end result?

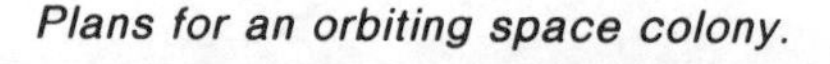

Plans for an orbiting space colony.

I raised these questions at the end of a lecture at the Smithsonian Institution. The audience mulled over the problem and could see the difficulty of moving from earth into space, even if it were technically possible. When I got home I emptied my slide tray from the lecture and found a folded wad of paper jammed into the last slot. I opened it. In neat handwriting it said: "This is your audio-visual operator. I am a computer expert. Microscopic printed circuits will be the genes of the future. A form of life created by humans and based on silicon chips will replace life on earth as mankind is destroyed by pollution. This artificial intelligence will be sent out to replicate and to conquer the universe. . . ."

Machines—mindless wanderers in a lifeless universe. I recognized the theme. The note referred to work done by the famous Professor John von Neumann, born in Budapest December 28, 1903, colleague of Einstein at Princeton, and an expert on chance and game theory. Thirty years ago, von Neumann put forward a mathematical proposition, a mind-product, something like Schwartzchild's black hole equations. He proved that it was possible to build a self-replicating, computer-controlled, voyaging robot. The whole thing would need to weigh 100 tons or more, but once designed and built, nothing would stop it. New robots would be made out in space from material dug (by robots) from the surface of small asteroids. Damage by meteorite hits and wear and tear on the system could be fixed under prearranged computer instructions. A single von Neumann machine could make 10 more; 10 could make 100; 100 could make 1,000. . . . Ultimately the Milky Way galaxy would be populated by these machines, colonized by mechanical clones.

Granting that Professor von Neumann's equations are right—and he was acclaimed a general scientific prodigy in his own time—one wonders at the idea of busy, spreading, nonhuman robots. Rabbi Löw of Prague made the Golem out of clay and didn't like the result; the Sorcerer's Apprentice could not control the runaway bucket! Surely to create a procreating silicon chip with the power to dominate the cosmos would be a hollow ending for the present mindstep. Sapiens would have been entirely the wrong name to have chosen for our species.

FOURTEEN
Future Mindsteps

I have singled out 4 astronomical mindsteps in the long sweep of history, and a zero-level threshold before that. Mindsteps can be likened to a staircase, where the new idea is the riser and the adjustment that follows is the tread. Or they can be looked on as steps forward in a long journey of understanding. There were others, of course, but I have selected these 5 leaps in understanding as the big ones, the ones with the greatest human impact. Galileo's telescopic discoveries could be included as a mindstep perhaps, but those events do not seem to me to be of the same magnitude as 0, 1, 2, 3, and 4. Copernicus reduced the earth to an ordinary planet and turned the stars into suns, but Galileo's discoveries only added emphasis and proof positive of the new viewpoint. If pressed further on this I think I would class Galileo's work as a substep—number 3.1, a part of the on-going Copernican revolution. Similarly Isaac Newton's discovery of the law of gravity would be 3.2. Some historians give greater significance to Kepler. His discovery of the elliptical orbits was far closer to the truth than the epicycles of Ptolemy and Copernicus, and the 3 planetary laws were vital precursors to Newton's work. If so, then Galileo moves to 3.2, and Kepler becomes 3.1. Then there is the question of where Albert Einstein's theory of relativity fits—is it a step or a substep? Personally I place relativity more in the field of physics than astronomy, more in the Age of Revolution than the Age of Space. True, general relativity predicts black holes and explains the redshift of distant galaxies, but even these astronomical concepts do not affect humanity with the major impact of a mindstep. Einstein's theory dealt with energy, photons and fast-moving things, not with life, and it is this human perspective that is so basic to a mindstep. These are all important points, but I will leave them open for debate.

Nor do I equate mindsteps with what others have called "paradigms," or "new world views." A mindstep is an example of both of these, but is more dramatic and irreversible. The word "paradigm" now covers almost any pattern of thought or collective policy of a group, and world views can be moral or political and can change in cycles. With the cosmic mindsteps I see no way of returning to stages 3, 2, 1, or 0, unless the world's humanity loses its sapience, or its store of accumulated knowledge.

No, the mindsteps do not go in cycles. Each one takes the collective mind closer to reality, one stage further along in its understanding of the relation of humans to the cosmos. There was nothing to stop the heliocentric theory arising once the base of Greek science and philosophy was laid. Aristarchus had already proposed it in 270 B.C. A Copernicus, if there had been one, could have started from there. But for some reason the human mind of the 3rd century B.C. could not take it. The time had not yet come. Aristarchus was denounced for impiety by the Stoics, and his idea withered. The notion of a moving earth was a shock even in the Middle Ages. The senses argued against it. Ptolemy in his time knew there was no feeling of motion, no wind constantly blowing in the hair, no sense of dizziness. Nor did the idea of the earth as a small planet fit in with the age-old myths and legends of goddess Ge, nor the "natural" place of mortals at the center of

The technologies of mindsteps 1, 3 and 4 on a British postage stamp.

the universe. Yet despite the inertia, even if it had to be via Ptolemy, via mindstep 2, the heliocentric theory had to come.

The mindsteps are cumulative—each one overlays the others in the collective mind. As in climbing a stepped pyramid, steps 1 and 2 are dimly remembered on reaching step 3. The collective mind still carries with it the shock of that great event when earth and sky were distinguished, the enjoyment of myths and legends to explain what can be seen overhead, and a fascination for unidentifiable moving points of light. Even now, in a grade-school class, there will be some doubters when the teacher says the earth is a spinning ball going around the sun.

Are there future mindsteps? Must we go through them? Could we jump to the end of the sequence suddenly, say tomorrow? History so far has shown a measured pattern. There needs to be a long period of sinking in, of absorbing the step. It seems to take time for the world to build up a mental attitude, a responsive mode. If Copernicus had lived in 500 A.D., I do not think his new theory would have survived. Clever and receptive though he was, the atmosphere of learning in 500 A.D. might not have put the idea in his head. It is only a speculation, but there may indeed have been a person equivalent to Copernicus then, or even later, but he or she was locked into mindstep 2 as was the rest of the Western world. Again it is a speculation, but I think a Copernicus living in the ancient Orient would never have gotten to the heliocentric theory because of the influence of Yin and Yang. No, it seems that there is no quick or sudden advance; the collective mind moves slowly as year succeeds year.

The waiting period between the mindsteps is getting shorter. One can't help noticing the acceleration. Mindstep 2 can be placed historically at about 150 A.D., 3 at 1543 A.D., and 4 somewhere in the 20th century—say 1926, the date when Robert Goddard fulfilled his dream by launching the first liquid-fuel rocket. Mindstep 1 cannot be fixed more accurately than to within a few millennia. It shows up in the first written records around 300 B.C., and may have begun sometime before then. Mindstep 0, the awakening of concern for things of the cosmos, could have been as far back as 35,000 B.C., or even earlier.

Each mindstep heralds in an Age—that interval of time for absorption and diffusion before the next mindstep comes along. The Age of Chaos, the alpha era, was between mindstep 0 and 1. The cosmos was seen, registered in the mind, but was unex-

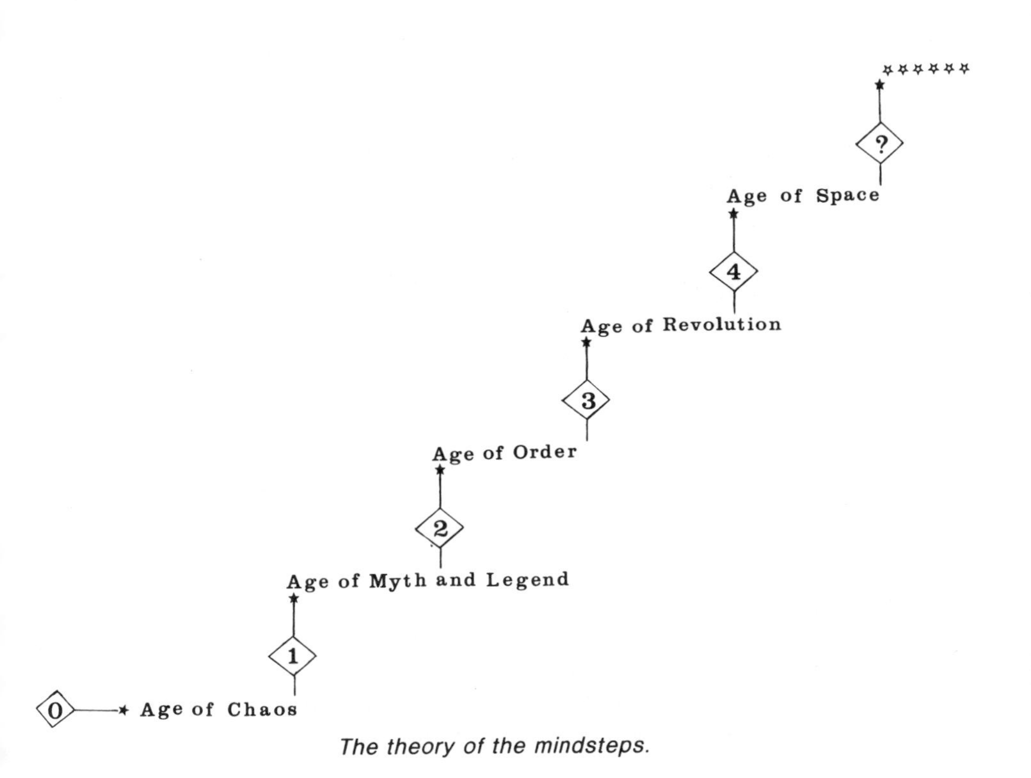

The theory of the mindsteps.

plained. The Age of Chaos lasted 32,000 years according to this tentative chronology. This was followed by the Age of Myth and Legend, when the sun, moon, stars and planets were explained by stories. It lasted 3,150 years and was brought to a close by astronomer-mathematician Ptolemy. Next came the Age of Order, which lasted 1,393 years and was terminated by Copernicus. This mindstep (3) brought in the Age of Revolution, 383 years long, which in turn led to the Age of Space. In summary:

Mindstep 0, 35,000 B.C., began the . . .
Age of Chaos (32,000 years)

Mindstep 1, 3,000 B.C., began the . . .
Age of Myth and Legend (3,150 years)

Mindstep 2, 150 A.D., began the . . .
Age of Order (1,393 years)

Mindstep 3, 1543 A.D., began the . . .
Age of Revolution (383 years)

Mindstep 4, 1926 A.D., began the . . .
Age of Space

How long does it take for a new cosmo-human perspective to come along? Does humanity need to gestate before a new mindstep can be brought forth? Just as the developing embryo must grow arms and legs to come to full term, the world's mind over the centuries must grow appendages. In ordered sequence these mechanical appendages of the mind have been the paint brush, the pen, mathematics, printing, and now the computer.

According to these very rough figures there is an acceleration; the waiting period gets shorter as the mindsteps slip by. The Age of Myth and Legend was about one-tenth as long as the Age of Chaos which preceded it; there was a 10-fold acceleration in getting to mindstep 2. But subsequent waiting periods show approximately a 3-fold acceleration, which is less violent. Even so, the interval between mindsteps has become very short—the "wait" for mindstep 4 was only 383 years, more or less, depending on what specific historic event one chooses to fix the dates.

After I had given the last lecture in the Smithsonian series, I read all the course evaluation sheets. Several students had asked

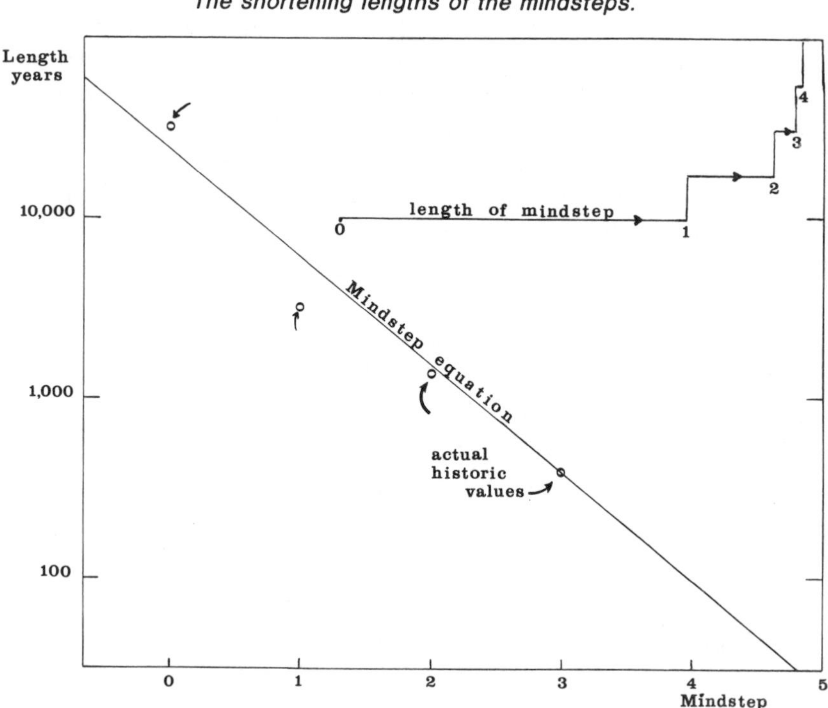

The shortening lengths of the mindsteps.

whether or not there was an algorithm for calculating mindstep dates on a pocket calculator. I had not done so in the course because mathematically there would be no meaning to fitting the numbers into an equation. History is a sociological phenomena and can't be put into a straightjacket. The cosmos might march on numbers, but human events do not. Yet the question was an interesting mathematical challenge, an exercise in numbers.

If one starts with my last two mindstep dates and assumes a constant acceleration of 4, then the various events in history can be forced to a "mindstep equation."

$$\text{Mindstep M} = -30{,}629 + 98{,}048(1/4 + 1/4^2 + 1/4^3 + \cdots + 1/4^M)$$

The first term is really mindstep 0 and has been assigned an ASTRODATE of −30,629, or 30,630 B.C. Each new term that is added gives the mathematical date for the next step:

Mindstep	*Astrodate*
0	−30,629
1	−6,117
2	11
3	1543
4	1926
5	2021
6	2045
7	2051
End of series	2053 A.D.

These values lie on a neat, deterministic, straight line. With or without human endeavor, says the computer, mindstep 5 is due in 2021 A.D.

Continuing purely as a mathematical exercise, one can with algebraic changes rewrite the equation to take care of fractions of a mindstep. The end date to the series is 2053.66, and for the general value of the mindstep, M:

$$\text{Mindstep (M)} = 2053.7 - 32{,}682.7/4^M$$

Putting in the values for decimal substeps, mindstep 3.1 comes out to be 1609 A.D., the year of Galileo's first telescopic discoveries: an M-value of 3.2 gives the year 1666 when Newton, in his own words, "began to think of gravity extending to the orb of the moon"; and 3.9 gives 1907, the discovery of relativity. (Einstein said: "My first thought on the theory of relativity was in 1907.

The idea occurred suddenly.") Continuing with the substeps of mindstep 4, one finds that 4.1 was in 1942 (the development of the von Braun V-2 rocket at Peenemünde?); 4.2 was in 1956.9 (close to the launching of the first earth satellite, Sputnik, in 1957); and substep 4.3 was, according to the equation, 1969—the year when humans first walked on the moon. All this seems alarmingly coincidental! But the equation begins to flag a little with mindstep 4.4. Here the predicted year was 1980, and history does not seem to have left a substep in the cosmo-human story (because of delays, the inaugural flight of space shuttle Columbia did not take place until 1981). Also, important events like the first landing on Venus or the robot touching down on the surface of Mars do not seem to fit with any substeps . . . Caveat Emptor!

Mathematics aside, there is the fact that these important events *have* taken place during the long course of history. And the various dates *do* seem to indicate a trend in the timing. But there is no way to predict mathematically the humanistic future from the past. At the basic level we are dealing with human aspirations, human nature. The mindsteps, even though set at certain junctures in history, are expressions of sociological, humanistic change, beyond the realm of numbers. The 31st century B.C. is within the acceptable range of time for mindstep 0, and it is possible that the Age of Myth and Legend might extend back to the 7th, but the mathematical forced-fitting has its most serious problems at the other end of the line. It makes quite a difference if one assigns the commencement of the Space Age either to Jules Vernes in 1865 or to Sputnik I in 1957. Also, according to the numbers, either Ptolemy was born more than 100 years too late, or the honor of mindstep 2 must be given to some (unknown) predecessor in 11 A.D!

But numbers aside, and even if the first 4 mindsteps have indeed shown a trend, what of the tricks of fate? What if the mother of Copernicus had decided not to be a mother and had taken vows in a nunnery—with no Copernicus, would there have been a theory? Looking at the writings of his contemporaries, probably not, at least not precisely in the year 1543. No, a mindstep equation can only be accepted for what it is, a mathematical game, an astronomical hypothesis—at the most intriguing, at the least unprovable. As the authorities of the Inquisition said in 1616 when instructing the censors of Copernicus's book:

> If certain of his passages on the motion of the earth are not hypothetical, make them hypothetical; then they will not be against either the truth or the Holy Writ. On the contrary, in a certain sense they will be in agreement with them, on account of the false nature of suppositions, which the study of astronomy is accustomed to use as its special right.

No, the human future is not contained in an equation, not even by supposition. However, based on the trend, it would be reasonable for us to prepare for the eventuality of the next mindstep, though when and what it will be is entirely speculation.

So far, all the mindsteps have been associated with inventions for improving meme transfer—drawing, writing, printing, computers. It is those latter day silicon-chip devices that have expanded the power of the human mind and made the Space Age possible. The Industrial Revolution took the drudgery out of handwork; the Computer Revolution takes the drudgery out of brainwork. Computers, like the other inventions, connect with the memes. They are the making of mindstep 4. Perhaps we will not be able to move to mindstep 5 until another breakthrough in communication technology comes along.

If rationality is any guide, one might expect the 5th mindstep to be communication with extra-terrestrial intelligence. Maybe it could happen with radio and the technologies we have on the shelf, or maybe it will be necessary to wait for a new, exotic communications device. With radio, or that yet-to-be-invented invention, planet earth would be sharing the memes of extra-terrestrial civilizations. And, according to the mindstep equation, that event would be in the year 2021 A.D.!

Many people are ready for extra-terrestrial communication to happen as a matter of course. Movies, TV and science fiction novels use the theme relentlessly. Spielberg's 1982 box-office success established "E.T." as a pseudonym for "Extra-Terrestrials." Radio astronomers now look and listen for those signals. The first outward-going message has already been speeded on its way from the radio dish in Arecibo, Puerto Rico, and it will reach various likely stars during the next 100 years. It was a simple message in the form of a TV picture, and the scientists are sure those E.T.s will be able to figure out the T.I. (Terrestrial Intelligence). If they responded immediately, the return receipt could be back at the earth in the 2021 time frame! Alternatively Pioneer 10 went clear out of the solar system on June 13, 1983. E.T.s can capture it, so the argument goes, and they can take the spacecraft apart and figure out how smart earthlings are. They have been given some clues about earth and the shape of man and woman engraved on a special gold-plated aluminum plaque mounted on the antenna of the spacecraft.

But this approach may be a null-precursor. There are those who believe space is entirely materialistic, empty of life, and who call the SETI project a fruitless waste of effort. There are many reports of UFOs around the world, on the ground and in the air, and of human meetings with E.T.s—those so-called close encounters of the third kind. But none of these reports has been established beyond reasonable doubt, and until that happens we are dealing with conjecture, not the beginnings of a mindstep.

Some astronomers tend to be optimistic. Given a planet like earth, they say, life will inevitably develop. There are billions of stars; there must be millions with E.T. civilizations. Not so, say the biologists: earth-like planets may be plentiful, but life itself is rare. Intelligent life hangs on the tip of one thin branch of the tree of evolution, and that branch could have snapped anytime in geological history.

Leonard Ornstein, pathologist at the City University of New York, is one of the more powerful debaters. He says the evolutionary track leading to intelligence is fraught with chance and risk. The dice have been thrown in such a bizarre sequence, it could happen once, and once only. Planet earth is our only known example, and here all life forms use levo (left-handed) amino acids. None are dextro (right-handed). Why? There seems to be no reason why DNA chains and organic molecules could not be made from the mirror-image building blocks, but they are not. Nowhere in all the millions of species is there dextro-based life. The simplest explanation is that life formed in just one of the many primeval pools, and all life forms on earth are derived from that single seed. It was a unique event.

Now uniqueness is difficult to deal with in probability theory. If life was a unique event on earth, it might have been a unique event in the entire universe. The argument goes like this:

> I have a billion marbles in a bag. (Ignore the size of the bag; this is a mathematical proposition.) The first one I pick out is

labeled "A planet with life." What are the chances of picking out a twin? Strict probability says there may be no twin left behind in the bag. All the other marbles may turn out to be labeled "A planet with no life." So earthlings can't prove (or disprove) there are other earthlings out there from the single labeled marble. It is a matter of believing or disbelieving.

From bacteria to intelligence along the vertebrate branch is a difficult evolutionary track, says Ornstein, and much has to happen on the way. Development of the hand or something similar is probably a prerequisite, and so is the eye. Without the eye the cosmos cannot be seen; the species will never even get to mindstep 0. Now nature in the evolutionary process does not put much store in vision. Bacteria, plants, the majority of species, function efficiently without it. Fish that revert to the caves quickly adopt a state of blindness. It takes too much biological energy to maintain such a complicated organ as an eye. Use it or lose it, the Darwinian rule seems to be.

An eye is a rarity in evolution. The human eye is related to the fish eye, and was almost lost in the course of evolution. Of the two principal types, the octopus eye is normal, but the fish eye is not. In the octopus, all the nerves come out from the back of the retina in a bundle, like the cable in a TV camera. But in the fish (and human) eye, all the nerves pass in front of the retina, and then funnel out through the optic nerve. Result: the network of nerves gets in the way of the image. This change probably took place way back in a sea-creature ancestor where the light was reflected from the back of the eye and the retina had to be inverted. This may have been useful way back then, but it is no advantage to descendents like the fish and us. In this type of eye the light has to pass through the nerves; so to develop a clear, sharp picture, the fish eye has developed a remarkable transparent nerve fiber. This was a complicated correction of a developmental error, and it permitted the various species along that branch to evolve. Without the transparent nerve fiber, intelligence would not now be found at the tip of the branch. Species Homo would not be sapiens. The odds were stacked against it. Even with the advantage of an eye with the retina the right way round, intelligence failed to appear on the octopus evolutionary track.

"Biological supposition!" one can say, and the biology can be dismissed in the way astronomy has been dismissed before, but these considerations do raise doubts about the universal existence of intelligent life. Mindstep 5 as extra-terrestrial communication might never come. In the code name for project SETI (Search for Extra-Terrestrial Intelligence), only the TI may have meaning. We may be entirely alone.

That is why I consider the possibility of an E.T. aspect for mindstep 5 to be so important. If we make contact it will open unimaginable horizons. It will truly be a mindstep. The fact of the contact would traumatize the world, and the results would alter the course of history. If there is no contact, and if we convince ourselves that planet earth is the unique home of cosmic intelligence, then that fact would have equally traumatic ramifications, for the challenge would be to move out from the pod and populate the galaxies, not with von Neumann machines, but with humans. As Pogo might then say: "We have found the intelligence, and it is us!"

The when and what of the coming mindstep are anybody's guess. If the past millennia are anything to go on, we are groping toward it and are currently in the 5th waiting period. Our problem will be to distinguish precursors from the real thing. The step may be totally beyond our present ken; it may be a shock. These precursor ideas are now being hotly, sometimes acrimoniously, debated in the ivory tower and outside. Personally I feel the step will be in some way an extra-terrestrial one, and that tends to be the viewpoint of most astronomers. The 3 anthropic principles lead in that direction. Yet many others think that communication from space intelligence would be nothing short of a miracle. Miracle or not, it seems to me that the slow adjustment of humans to the cosmos, through cave pictures, legends and telescopic viewing, has been leading toward a contact. The conquering of space in our generation is a remarkable achievement that is only slowly being absorbed. Humans have their foot on the threshhold and there is no turning back.

In looking into the future we cannot ignore the past. From this first look at the mindsteps they appear to have certain things in common—a new and unfolding human perspective, related inventions in the area of memes and communications, and a long formulative waiting period before the next mindstep comes along. None of the mindsteps can be said to have been truly anticipated, and most were resisted at the early stages. In looking to the future we may equally be caught unawares. We may have to grapple with the presently inconceivable, with mind-stretching discoveries and concepts.

Mind-step	*Date*	*Beginning of the . . .*	*Related Invention*	*Cosmos-human Perspective*	*Size of Cosmos*
0	35,000 B.C.	Age of Chaos (Alpha era)	Drawing, painting, carving	Universe centered on the person. Earth was flat. Celestial objects were merged with earthly objects.	Limited to local environment.
	3,000 B.C.	Age of Myth and Legend	Writing	Universe centered on an underworld. Earth a flat disk beneath a celestial ceiling or dome. Earth and sky separated. Cosmic objects explained as gods.	Contained under the celestial ceiling.
2	150 A.D.	Age of Order	Mathematics	Universe centered on earth as a fixed, nonrotating globe. Cosmic objects moved in ordered circles and epicycles. A space established between earth and celestial sphere.	Contained within the celestial sphere.
3	1543 A.D.	Age of Revolution	Printing, telescope	Universe centered on sun. Earth a globe, spinning and moving in orbit.	Boundless
4	1926 A.D.	Age of Space	Liquid-fuel rocket, computer, radio, TV	Center of the universe unknown. Earth a spinning geoid, moving with sun and solar system around galact center. Humans travel in space.	Limited by red-shift horizon, 18 billion light years.
5	2021? A.D.	Age of Cosmic Connection?	A new technology break through?	Communication with extra-terrestrial civilizations?	?
6? 7?		(Omega era?)	?	Complete?	?

Suppose for a moment that, deterministically or not, the mindstep equation works. It fits mindsteps 3 and 4 exactly; it fits 2 fairly well; and 1 is not outside of the acceptable range of dates. Also, the beginning of the alpha era, mindstep 0, is not badly represented by the nominal date of 30,630 B.C. As a physical scientist I do not play around with numbers in an equation. My rough estimates in the table were arrived at before the equation was produced, nor are they anything more than estimates. Having made my first guess it is not good research policy to make arbitrary alterations. However, a numerical analyst could obtain a better fit by tinkering with the data a little bit more. If mindstep zero were ordained to be in −28,193, and the coefficient changed to 90,624, then mindstep 2 would fall in 127 A.D. (comfortably within the lifetime of Ptolemy), 3 would still be 1543 A.D., but the Space Age should have begun in 1897, and mindstep 5 would be 1985—halfway through that particular year! To tinker this way is purely a mathematical exercise, conjecture. The algorithm is in no way predictive.

Taken on its own, the world of numbers can play strange tricks, and I wondered as I looked at the mindstep equation how the numbers ended. It is a series of terms (steps) which comes to a stop, a grinding halt. There is an absolute mathematical limit, and no matter how many mindsteps are added on, as M goes to infinity one meets a time barrier. There is no way to break through. For the particular values of coefficients assumed, the limit is in the 21st century. There is room for about 2 more mindsteps, the 6th in the year 2045, and the 7th in 2051 A.D. If the alpha era was heralded by mindstep 0, the omega era will be reached, according to the quirks of the equation, precisely in 2053 A.D. What happens mathematically (as the pocket calculator will show) is that a time comes when the mindsteps fall thick and fast. The acceleration has so increased that between mindstep 7 and the limit, mindsteps come every month, then every day, hour and second until in the year 2053 they are flashing by at infinite speed. At that time the human mind will have absorbed all the possible mindsteps there are to absorb, and in a grand climax humanity will make a final adjustment to the greater cosmos. Destiny, in terms of the equation, will have arrived!

One can hear the refutation echoing forward from those 17th century theologians: “Hypothetical!” “False nature of suppositions which the study of astronomy is accustomed to use!” And rightly so; it is an equation, nothing more. The cosmos marches on numbers, but humanity moves by the spirit. The planets, stars and galaxies can be reduced to coefficients and digits, but the mind and human progress defy analysis.

At this point the expertise of an astronomer meets Lewis Carroll's Boojum, and softly and suddenly vanishes away. If mindstep 5 really comes, could mindstep 6 and 7 be far behind? Surely any steps beyond E.T. contact would be enormous. If mindstep 5 is the establishment of communications with a network of cosmic intelligence, we would most probably be sharing knowledge and experience which transcends our own. (Or we might receive the ultimate threat!) On the other hand, no signals at all would mean that the challenge of an empty universe lay before us. Either way, whatever the challenge, the rightful place of living objects in the cosmos would be established for all time. There would be that part of the cosmos made of inert material, the Yin, and that part possessing the creative life force, the intelligence, Yang.

One possibility for mindstep 6 might be an invention for traveling in time. This would be an amazing thing for humanity to absorb because we would be communicating with our future and our past. It raises logical dilemmas such as, "Would the treading on a certain butterfly in the year X-thousand B.C. alter the course of evolution?" Or, "Seeing our fate enacted so many years from now, could we do something to change that fate?" Fortunately, a breakthrough in time travel—time warps—is not expected from the mathematics of physical science theories. The type of time travel predicted by Albert Einstein's relativity theory is always associated with extremely high velocities, or super-intense gravity. The one presents technical difficulties of accelerating close to the speed of light; the other presents the problem of surviving an encounter with an object as lethal as a black hole. The theory of relativity might fit into the mindsteps as 3.9, but I don't think it will succeed in its present form in giving us a step 5 or 6. It may so happen that further progress comes from areas outside of the hard sciences.

Mindstep 6 could be proof, the demonstration of the existence of, something unseen, a new type of field connected with life, the existence of which is so persuasive that its general acceptance is inevitable. In religious contexts it could be identified with the world of the spirit. In this area there are scientific indicators which may or may not be precursors. There are the broad questions of telepathy, existence or not of a vital force, the existence of the soul, and continuation of life after death. Then there are the unmeasurable human emotions of love, friendship and hate, pleasure and pain. Are these all interlocking parts of a nonphysical cosmos? The atoms of iron and carbon in the body, made in the core of stars, will ultimately be returned to the cosmos, but thoughts and brain waves go without leaving a trace. These things cannot be seen through the microscope or radio telescope, yet they are things that humans are made of.

Currently these topics are at the forefront of research. At the lower primate level there is the 100th-monkey experiment where researchers claimed a threshold for collective knowledge. In a large colony of monkeys on an island near Japan, the trick of washing sweet potatoes was taught individual by individual. Then, after the knowledge had been gained by about 100, suddenly the whole colony seemed to know the trick, even those on the other side of the island. The scientists involved felt it could not be explained by a normal learning process.

At the higher primate level there is the phenomenon of getting a new idea—where does it come from? Or the feeling of having met up with something before—when? The 3-year-old American Indian girl knew about the phases of the moon before she saw them in her storybook. Hinduism is founded on the principle of reincarnation, as is that section of Buddhism that seeks out the new Dalai Lama in the child born at the instant of death of the old Dalai Lama. Medical researchers are now considering the phenomenon of near-death experiences (NDEs), in which persons resuscitated after being pronounced clinically dead have related death experiences. Generally the experience is of leaving the body, floating with no influence of gravity, moving in a long tunnel toward a light at the end and reunion with loved ones. Some dismiss this as nothing more than violent dreaming of the dying brain, but the research is continuing.

Some of the great minds in the natural sciences have pondered over the question of future mindsteps—not in terms of the idea of the stepped pyramid that I have presented here, but in terms of the ultimate limits of the explorations of intelligent inquiry. British physicist Sir Oliver Lodge vowed to prove there was a life after death by communicating across the barrier if he reached the other side. He said if there was an afterlife, and if a way to communicate could be found, he would do so. He would devote all his energy to the task, treating it as a scientific challenge. He died in 1940 at his home near Stonehenge, and left a sealed message which he said he would reveal to the Psychical Research Society based near Piccadilly, London. A research team there made itself ready to receive the message during the blitz period of World War II, and worked on and off for about a year. The results, so they concluded, were negative—Lodge had failed. Yet without knowing anything about the nature of the experiment, one medium "saw" a gray-haired man, another "heard" music, and still another said letters were floating about and recognized "Stonehenge H." When the envelope was finally opened it contained the letters of a five-finger musical scale, but one not in total agreement with the music that was "received." Professor Lodge's home was a few miles from Stonehenge, but he had never been involved in any research there.

Nobel prize-winner Sir John Eccles, famed for his work on the structure of the human brain, once wrote: "Is this present life all to finish in death . . . or is there some further transcendent experience of which we can know nothing until it comes?" His ideas were sharply criticized by others in the field, but he theorized that the brain was like a computer, organic and material, yet there was within it a soul of ethereal substance. "This whole cosmos is not just running on and running down for no meaning," he declared.

Max Planck, founder of the quantum theory, the German physicist whose son was executed after an assassination attempt on Hilter in 1944, looked for something at the top of the mindstep pyramid when he said: "Religion and natural science are fighting a joint battle in an incessant, never-relaxing crusade against skepticism and against dogmatism, against disbelief and against superstition, and the rallying cry in this crusade has always been and always will be 'On to God.'"

In suggesting mindstep 6 as the Age of the Spirit, would there be a barrier to reaching it, would a new technological invention be needed before this step could happen? I am speaking of this, of course, in the strict connotation of a mindstep, the spreading and acceptance of a secure knowledge worldwide. It would have to be an unexpected development that would transform a belief into a proof, or something approaching a certainty—a substantive advance for human understanding.

If the mindsteps were to continue this acceleration toward the mindstep limit, could a mindstep 7 follow a 6? Mindstep 0, the beginning, would be the alpha era, and 7 would be the omega. From the broad humanistic perspective, after communicating with other civilizations (5), and a spiritual domain (6), the ultimate step (7) could only involve tangible contact with a superintelligence, the Omnipotent. Not so much the "Black Cloud" in the experimental novel by Fred Hoyle, because that cloud of intelligence was not unique, but contact with the universal mind talked of by Aristotle and embodied in world religions—proof replacing belief. Here we come to the transcendental barrier long recognized through the Ages.

We are at the close of the 20th century and our fellow humans have considered this question since the time of the Pharoahs and before. In our time atheism is the official belief of one of the world's super-powers, and still other countries are agnostic. The God of one religion is not easily equated with the God of another. Such differences, even today, can take sovereign states to war, shedding blood, and a collective understanding does not seem possible. There seems to be nothing in place in the late 20th century with the magnitude or momentum of a 7th mindstep. Only time will tell whether the various religious beliefs are nothing more than precursors, or whether they are going to be the door which will lead the earth to its future mindstep.

But the astronomer's world of stars and galaxies is out there, and we have been privileged to see it and study it. Knowing more about the cosmos has lifted us from chaos and freed us from the dogma and naïvete of those older mindsteps. Our present mindstep, 4, itself in its beginning, has only just begun to give us startling new wonders to exchange for the old.

Appendix

Computer Programs

The astronomical equations can be simplified for use on a pocket calculator or home computer. These programs give the positions of the planets, phases of the moon and other quantities. The program language is GS/H® (General Scientific/Home), and is designed to be entirely independent of the type of computer used. It can be translated into COBOL, FORTRAN, or whatever languages are required of future models of home computers. For a nonscientific calculator with limited storage in the memory, a pencil and pad will be needed as an extra. Functions such as sine and cosine can be looked up in trigonometric tables and keyed in; numbers needed to be stored in memories can be written down and then keyed in when required. But a regular scientific calculator will have all of the functions called for. The average home computer will have more than sufficient memory, and the various programs can be permanently encoded and called up as subroutines.

In developing these programs I have intended them to be as simple as possible in the beginning routines so that a nonmathematician can see the principles involved. In the more complex programs I have provided only the equations, and leave the programming to the reader. Many users may prefer a different logic in placing the equations on the machine, though, as a caveat, it must be pointed out that in departure from the given format, "bugs" may appear. All the programs given here were tested by students in my course at the Smithsonian Institution, and we all hope that no bugs remain in the given formulations. I would appreciate comments from readers as to improvements, additions and changes.

Most of the math terms are standard and recognizable. The following ones are specifically defined:

Int: = Integer, take the number before the decimal point.
Example: Int: 41.76 = 41

F: = Fraction, take the digits after the decimal point.
Example: F: 41.76 = 0.76, and F: −41.76 = −0.76

Pos. F: = Positive fraction, take the digits after the decimal point; if positive, take the value; if negative, add 1 and take the value.
Example: Pos. F: +41.76 = 0.76, and Pos. F: −41.76 = 0.24

exp = exponential, take the exponential.
Example: exp(2) = e^2, where e = 2.718

$\underline{\qquad\qquad \underline{\qquad\qquad\qquad}}$ are mathematical brackets.

Example: Int: $\underline{2 \times \underline{1 + 1.1}}$ = Int: (2(1 + 1.1))

arcsin = the angle whose sine is . . .
Example: arcsin: 0.5 = 30 degrees

+/− = reverse the sign.
Example: +/−: −24 = 24

Abs: = the absolute value of a number.
Example: Abs: −1.2 = 1.2, and abs: +41.76 = 41.76

logn = the natural logarithm of a number.
Example: logn: 4.0 = 1.386

Program NOONALT

This program gives the altitude in degrees of a celestial object above the southern horizon when it is due south at noon. The inputs to the program are geographical latitude of the site, LATITUDE, and the astronomical DECLINATION of the object.

NOONALT = 90 − LATITUDE + DECLINATION

(If NOONALT is greater than 90, then value is 180 − NOONALT)

The answer is accurate to within a fraction of a degree. Refraction in the earth's atmosphere will make the value slightly higher.

Program	Example 1	Example 2
NOONALT	DECLINATION −24 LATITUDE +30 (North)	DECLINATION +62 LATITUDE +30
Enter 90	90	90
−		
Enter LATITUDE	30	30
=	60	60
+		
Enter DECLINATION	−24	62
=	36	122
Display is NOONALT	36	122
If greater than 90	no	yes
Reverse sign	↓	−122
+ 180		180
=		58
Display is NOONALT	36	58
End	. . .	. . .

Program SUNMAX

SUNMAX is the maximum declination of the sun in its annual journey around the zodiac. At midwinter the sun's declination is −SUNMAX, at midsummer, +SUNMAX. It is also the value of obliquity, the tilt angle of the earth's axis to the orbital plane.

Input is YEAR A.D., Old Style or New Style calendar.

Note, for years B.C.,

YEAR = − (year B.C. − 1)

SUNMAX = 23.7 − 0.00013 × YEAR

The program accuracy is one-tenth degree or better between 2000 B.C. and 2000 A.D.

(continued on next page)

Program	Example 1	Example 2	Program	Example 1	Example 2
SUNMAX	2000 B.C.	2001 A.D.	0.00013	0.00013	0.00013
Enter year	2000	2001	=	−0.259	0.260
If year is B.C.	yes	no	Reverse sign	0.259	−0.260
− 1	1	↓	+ 23.7	23.7	23.7
=	1999		=	23.9	23.4
Reverse sign	−1999		Display is		
Continue	↓	↓	SUNMAX	23.9	23.4
If YEAR is A.D.	↓	yes	End	• • •	• • •
×					
. . . continue in next column					

Program MOONMAX

MOONMAX is the maximum declination reached by the moon during a given year. The midwinter full moon is at declination +MOONMAX, and the midsummer full moon is at declination −MOONMAX. The evening crescent is close to +MOONMAX at midsummer, and −MOONMAX at midwinter.

Input is YEAR A.D. in either the Old Style Julian or New Style Gregorian calendar.

Note, for years B.C.,

$$\text{YEAR} = -(\text{year B.C.} - 1)$$

$$\text{MOONMAX} = 24 + 5 \times \text{Cosine}\left(360 \times \text{Positive F: } \underline{\underline{\text{YEAR} - 14.9/18.614}}\right)$$

Program is accurate to within a fraction of a degree.

Program	Example 1	Example 2	Program	Example 1	Example 2
MOONMAX	1271 B.C.	1997 A.D.	× 360	360	360
Enter YEAR	1271	1997	=	349.67	174.37
If year is B.C.	yes	no	Cosine	0.983	−0.995
−1	1	↓	× 5	5	5
=	1270		=	4.918	−4.975
+/−	−1270		+24	24	24
Continue	↓	↓	=	28.9	19.0
If Year is A.D.	↓	yes	Display is		
−14.9	14.9	14.9	MOONMAX...	28.9 . . .	19.0
=	−1284.9	1982.1		(year of	(year of
divide by				high moon)	low moon)
18.614	18.614	18.614	End	• • •	• • •
=	−69.028	106.484			
Fraction	−0.028	0.484			
If display					
is negative	yes	no			
+1	1	↓			
=	0.971				
Continue	↓				
If display	↓	↓			
is positive	yes	yes			
. . . continue in next column					

Program AZIMUTH

This program gives the azimuth A measured east of north for an object such as a star or the center of the disc of the sun or moon when at an altitude S. The program can be used to give azimuth when rising or setting. The equations are those of the STONEHENGE program, with q = 0. The parallax, p, for the sun is 0.002; for the moon, 0.951, when on the horizon. For a star, p = 0.

Program	Example	Program	Example	Program	Example
AZIMUTH	Heel stone	store in M 8		recall M 7	0.0018
enter latitude	51.178	×		=	0.403
store in M 1	M 1 = L	recall M 7	0.474	store in M 7	
enter dobj	23.906	=	0.469	recall M 0	0.132
store in M 2	M 2 = dobj	store M 9	M 9	cos	0.999
enter p	0.002	recall M 6	0.6	store in M 3	
store in M 4	M 4 = p	−		recall M 1	51.178
enter H	102	recall M 9	0.469	cos	0.626
store in M 5	M 5 = H	=	0.130	×	
enter S	0.6	+		recall M 3	0.999
store in M 6	M 6 = S	recall M 4	0.002	=	0.626
recall M 6	0.6	=	0.132	divide by	
divide by 3.0	3.0	store in M 0	M 0 = G	recall M 7	0.403
=	0.2	sine	0.0023	=	1.553
+/−	−0.2	store in M 7		1/x (recprcl)	0.643
exp (exponentl)	0.818	recall M 1	51.178	arccosine	49.944
× 0.58	0.58	sine	0.779	display is A for	
=	0.474	×		object rising . . .	49.944
store in M 7		recall M 7	0.0023	+/−	−49.944
recall M 5	102	=	0.0018	+	
divide by 8400	8400	store in M 7		360	360
=	0.012	recall M 2	23.906	=	310.055
+/−	−0.012	sine	0.405	display is A for	
exp (exponentl)	0.987	−		object setting	310.055
. . . continue in next column		. . . continue in next column		End	. . .

Program SUNDEC

SUNDEC, the declination angle of the center of the disc of the sun in degrees above or below the celestial equator is given by:

SUNDEC = arcsin (sin LONGITUDE × sin SUNMAX)

The program requires the input supplied by two other programs and ASTRODATE. For results where the accuracy is required to 0.1 degree, use LONGITUDE as a subroutine. For accuracies of about 0.5 degrees use AGE as a subroutine for sun's longitude. For approximate results the inputs may be estimated—for example, in the present era SUNMAX = 23.5, the vernal equinox is LONGITUDE = 0, and the solstices are 90 and 270 degrees.

(continued on next page)

Program	Example 1	Example 2
SUNDEC	vernal equinox	summer solstice
enter SUNMAX	23.5	23.5
sin	0.398	0.398
store M 1		
enter LONGITUDE	0	90.0
sin	0	1.000
×		
recall M 1	0.398	0.398
=	0	0.398
arcsin	0	+23.5
display is SUNDEC	0	+23.5
end	...	...

Program DAYS (Programmable)

DAYS is the number of days and the fraction of a day that have elapsed since the beginning of the year, taken as 00 hours U.T., midnight on January 1. Input is date and time. The program can be set up directly or entered as a subroutine. It uses program LEAP. The program is accurate.

FRACTION = HOURS U.T./24.

$$\text{DAYS} = \text{DAY} - 1 + \text{FRACTION} + \text{Int: }\underline{\text{MONTH} \times 30.6 - 32.3} \times (1 - \text{Int: }\underline{\underline{12 - \text{MONTH}}/10})$$

$$+\ 31(\text{MONTH} - 1) \times \text{Int: }\underline{\underline{12 - \text{MONTH}}/10}$$

$$+\ \text{LEAP} \times (1 - \text{Int: }\underline{\underline{12 - \text{MONTH}}/10})$$

Program	Example 1	Example 2	Example 3
DAYS	585 B.C. May 28 12 U.T.	5 A.D. March 28 15 U.T.	2001 A.D. Jan. 1 00 U.T.
enter MONTH	5	3	1
store in M 1			
+/−	−5	−3	−1
+12	12	12	12
=	7	9	11
divide by 10	10	10	10
=	0.7	0.9	1.1
integer	0	0	1
store in M 2			
+/−	−0	−0	−1
+1	1	1	1
=	1	1	0
store in M 3			
recall M 1	5	3	1

(continued on next page)

Program	Example 1	Example 2	Example 3
× 30.6	30.6	30.6	30.6
=	153.0	91.8	30.6
−32.3	32.3	32.3	32.3
=	120.7	59.5	−1.7
integer	120	59	−1
×			
recall M 3	1	1	0
=	120	59	0
store in M 4			
recall M 1	5	3	1
−1	1	1	1
=	4	2	0
× 31			
=	124	62	0
×			
recall M 2	0	0	1
=	0	0	0
store in M 5			
enter DAY	28	28	1
−1	1	1	1
=	27	27	0
+			
recall M 4	120	59	0
=	147	86	0
store in M 6			
enter LEAP			
from			
subroutine	1	0	0
×			
recall M 3	1	1	0
=	1	0	0
+			
recall M 6	147	86	0
=	148	86	0
store in M 7			
enter HOUR	12	15	0
divide by 24	24	24	24
=	0.5	0.6	0
+			
recall M 7	148	86	0
=	148.5	86.6	0.0
display is . . .			
DAYS . . .	148.5	86.6	0.0
End	• • •	• • •	• • •

Program DAYS (shortcut)

DAYS is the number of days and fraction of a day that have elapsed since the beginning of the year, 00 hours midnight on January 1.

FRACTION = HOURS U.T. / 24

DAYS = DAY − 1 + A,

where DAY is the day of the month as given by the date, and A is given by the table:

Table for Value of A

Month	*Regular Year*	*Leap Year*
1	0	0
2	31	31
3	59	60
4	90	91
5	120	121
6	151	152
7	181	182
8	212	213
9	243	244
10	273	274
11	304	305
12	334	335

Leap years can be determined from program LEAP, or the "division by 4 rule." Because of the inspection of the table values, this shortcut is not readily programmable.

Program	Example 1	Example 2
DAYS	May 28.5, 585 B.C.	March 28.6, 5 A.D.
enter DAY	28.5	28.6
−1	1	1
=	27.5	27.6
+		
enter A from table	121	59
=	148.5	86.6
displays is DAYS . . .	148.5	86.6
End	• • •	• • •

Program LEAP

The first step in dealing with a date is to find out whether or not it was in a leap year. Program LEAP tells whether or not a year has a length of 366 days with an extra added on February 29.

For a leap year, LEAP = 1. For a normal year, LEAP = 0.

The quantity YEAR is the number of the year in the A.D. system. It can be positive or negative.

Note: For years B.C.,

YEAR = − (year B.C. − 1)

For the old style Julian calendar as used in history, the year was leap if the year number was divisible by 4. Thus,

LEAP = Int: $\overline{1 - \text{F: Abs: YEAR/4}}$

For the new style Gregorian calendar as used today, the year is leap if the year number is divisible by 4 and, if the year is a century, also by 400. Thus,

LEAP = Int: $\overline{(1 - \text{F: Abs: } \underline{\text{YEAR/4}} - \text{F: } \underline{\text{1/4 Int: } \underline{\text{YEAR/100}}} \times \text{Int: } \underline{1 - \text{F: } \underline{\text{YEAR/100}}})}$

As a shortcut, leap years can be found by straight mental arithmetic, using the "divisible" rules above, but program LEAP makes the problem a push-button operation. On a home computer, LEAP can be filed as a subroutine.

Program	Example 1	Example 2
LEAP	2000 A.D.	2000 B.C.
	New Style	Old Style
enter year	2,000	2,000
Is year A.D.?	yes	no
−1		1
=		1,999
+/−	↓	−1,999
store in M 1	2,000	−1,999
Is the date		
old style?	no	yes
enter 0		0
store in M 3	↓	M 3 = 0
divide by 100	100	(old style)
=	20	
fraction	0	
+/−	−0	
+1	1	
=	1	
integer	1	
store in M 2		
recall M 1	2,000	
divide by 100	100	
=	20	
integer	20	
divide by 4	4	
=	5	
...continue in next column		
fraction	0	
×		
recall M 2	1	
=	0	
store in M 3	M 3 = 0	↓
recall M 1	2,000	−1,999
divide by 4	4	4
=	500	−499.75
absolute	500	499.75
fraction	0	0.75
+		
recall M 3	0	0
=	0	0.75
+/−	−0	−0.75
+ 1	1	1
=	1	0.25
integer	1	0
display is		
LEAP . . .	1	0
End	• • •	• • •
	(2000 A.D.	(2000 B.C.
	was a leap	was not a
	year)	leap year)

Program ASTRODATE for the Old Style Calendar (OS)

ASTRODATE is the flow of time measured from a given starting instant. It is measured in mean Julian years of length 365.25 days. The start is taken as midnight on January 1, Zero A.D. in the Old Style Julian calendar, OS. It is the commencement of the day, January 1, 00 hours Universal Time (U.T.), 0 A.D. YEAR is the year number in the A.D. system. Years B.C. are negative, and, unfortunately, historians did not allow for a year "zero B.C."

For years B.C.,

YEAR = − (year B.C. − 1)

The input is YEAR and DAYS, where DAYS is the fraction of the year. DAYS is a subroutine for computing the fraction of the year based on the calendric date and the hour U.T. The program is accurate.

ASTRODATE = 1/365.25 (365 × YEAR + Int: $\underline{\underline{\text{YEAR + 4711}}}$/4 + DAYS − 1177)

Note: ASTRODATE can also be found from the so-called Julian Day Number, JD:

ASTRODATE = 1/365.25 (JD − 1721057.5)

N.B. care must be taken in using the JD system because the zero of the day-count starts at noon, U.T., or with modern precision in defining time, noon Ephemeris Time. The Julian Day Number system was defined to start at noon U.T. for January 1, 4713 B.C., JD = 0.0. In the ASTRODATE system, which follows the modern convention of starting the day at 00 hours U.T. (or in very precise work, E.T.), this 12-hour difference must be allowed for. For example, JD for noon January 1, 1900 A.D. is 2415021, and so ASTRODATE for January 1.0, 1900 A.D. is

ASTRODATE = 1/365.25 (2415020.5 − 1721057.5) = 1899.967

Program	Example 1	Example 2
ASTRODATE	May 28.5 585 B.C.	March 28.6 5 A.D.
enter year	585	5
is year A.D.?	no	yes
−1	−1	↓
=	584	
+/−	−584	
store in M 1	−584	5
× 365	365	365
=	−213160	1825
store in M 2		
recall M 1	−584	5
+4711	4711	4711
=		
divide by 4	4	4
=	1031.75	1179.0
integer	1031	1179

(continued on next page)

Program	Example 1	Example 2
add to M 2		
enter DAYS from subroutine		
DAYS	148.5	86.6
add to M 2		
recall M 2	−211980.5	3090.6
−1177	1177	1177
=	−213157.5	1913.6
divide by 365.25		
=	−583.59342	5.2391512
display is		
ASTRODATE . . .	−583.5934	5.2391
End	• • •	• • •

As a shortcut, accurate to 0.003, i.e., ± 1 day, ASTRODATE can be found directly from OS dates, using the year number and the number of days which have elapsed since January 1. Some calendars give the date as the "Nth" day of the year, and then DAYS = N − 1.

Short Cut: ASTRODATE = YEAR + DAYS/365.

Example: March 28, 1400 hours U.T., 5 A.D.

ASTRODATE = 5 + 86.6/365 = 5.237).

Program ASTRODATE for the New Style Calendar (NS)

ASTRODATE is the flow of time measured from a given starting instant. It is measured in mean Julian years of length 365.25 days. The start is taken as midnight on January 1, Zero A.D. in the Old Style Julian calendar, OS. It is the commencement of the day, January 1, 00 hours U.T., YEAR 0 A.D. YEAR is the year number in the A.D. system. The program below allows for the difference in length of the new style, Gregorian year (365.2425 days), and allows for the adjustment when days were dropped from the calendar in converting from Old Style to New Style. The input to the program is YEAR and DAYS, where DAYS is the fraction of the year. Program DAYS is a subroutine for computing the fraction of the year based on the calendric date and the hour U.T. The program is accurate.

$$\text{ASTRODATE} = 1/365.25\ (365.2425 \times \text{NEWSTYLE} + 2.0)$$

$$\text{NEWSTYLE} = 1/365.2425\ (365 \times \underline{\text{YEAR} - 1600} + \text{Int: } \underline{\underline{\text{YEAR} - 1597/4}}$$
$$- \text{Int: } \underline{\underline{\text{YEAR} - 1601}/100} + \text{Int: } 1/4 \text{ Int: } \underline{\underline{\underline{\text{YEAR} - 1601} / 100}} + \text{DAYS}) + 1600$$

(continued on next page)

Program	Example	Program	Example
ASTRODATE for	1900 A.D.	add to M 2	
New Style date	Jan. 1	recall M 3	2
	00 U.T.	divide by 4	4
enter YEAR	1900	=	0.5
store in M 1		integer	0
−1600	1600	add to M 2	
=	300	enter DAYS from	
X 365	365	subroutine DAYS	0
=	109500	add to M 2	
store M 2		recall M 2	109573.0
recall M 1	1900	divide by	
−1597	1597	365.2425	365.2425
=	303	=	300.0006
divide by 4	4	+1600	1600
=	75.75	=	1900.0006
integer	75	X 365.2425	365.2425
add to M 2		=	693961
recall M 1	1900	+2	2
−1601	1601	=	693963
=	299	divide by	
divide by 100	100	365.25	365.25
=	2.99	=	1899.967
integer	2	display is	
store in M 3		ASTRODATE . . .	1899.967
+/−	−2	end	• • •
. . . continue in next column			

As a shortcut, accurate to 0.003, i.e. ± 1 day; ASTRODATE can be found directly from NS dates, using the year number and the number of days which have elapsed since January 1. Some calendars give the date as the "Nth" day of the year, and then DAYS = N − 1.

Shortcut: ASTRODATE = 0.9999794(YEAR + DAYS/365) + 0.005

Example: January 1, 00 hours U.T., 1900 A.D. NS.

ASTRODATE = 0.9999794(1900 + 0/365) + 0.005 = 1899.965

Program AGE

Program AGE tells how far a cycle has progressed. For the five planets and the moon, AGE is the number of days checked off in the celestial body's journey in the zodiac, starting with a line-up at conjunction. LENGTH is the length of the cycle, which, for the planets and moon, is called the synodic period. AGE also gives the stage of the high-low 18.6-year cycle of the moon with output in years, and the longitude of the sun in the ecliptic measured from the vernal equinox in degrees. Accuracy can be increased by adding terms correcting for the elliptical orbits as in program LONGITUDE.

AGE = LENGTH × Positive Fraction: ASTRODATE − START/PERIOD

Values

	LENGTH	PERIOD	START	Accuracy
Mercury	116 days	0.3172554	0.333	(10)
Venus	584 days	1.598690	0.442	(5)
Mars	780 days	2.135350	1.721	(30)
Jupiter	399 days	1.092085	0.703	(7)
Saturn	378 days	1.035160	0.410	(11)
Moon	29.5 days	0.08085034	0.4694	(1)
Moon cycle	18.61 years	18.614	14.9	(0.1)
Sun longitude	360 degrees	0.9999786	0.223	(3)

Cosmic Events According to AGE

Synodic event	Mercury	Venus
Superior conjunction	0	0
Appears as evening star	20	39
Maximum evening elongation	36	221
Maximum brilliancy	42	257
Fades as evening star	48	285
Inferior conjunction	58	292
Appears as morning star	68	299
Maximum brilliancy	74	327
Maximum morning elongation	80	362
Fades as morning star	104	545
Superior conjunction	116	584

Synodic event	Mars	Jupiter	Saturn
Superior conjunction	0	0	0
Appears as morning star	65	20	20
Begins retrograde motion	353	140	125
Opposition	390	200	189
Ends retrograde motion	427	260	253
Fades as evening star	725	379	358
Superior conjunction	780	399	378

Moon phase	AGE	Season	Sun's LONGITUDE
New (conjunction)	0	*Vernal equinox*	0
Appears, evening crescent	1	*Summer solstice*	90
First quarter	7	*Autumnal equinox*	180
Full	15	*Winter solstice*	270
Last quarter	22	*Vernal equinox*	0
Fades, morning crescent	28		
New (conjunction)	29.5		

(continued on next page)

Program	Example 1	Example 2	Example 3
AGE	Jupiter 1493 A.D. OS Jan. 15.0	Moon 585 B.C. OS May 28.5	Sun's longitude 2001 A.D. NS Jan. 1.0
enter ASTRODATE	1493.038	−583.5934	2000.966
−			
enter START	0.703	0.4694	0.223
=	1492.335	−584.0628	2000.743
divide by . . .			
enter PERIOD	1.092085	0.08085034	0.9999786
=	1366.500	−7223.9993	2000.785
fraction	0.500	−0.9993	0.785
is fraction negative? . . .	no	yes	no
+1	↓	1	↓
=		0.0006	
continue			
×			
enter LENGTH	399	29.5	360
=	199.8	0.0	282.8
display is AGE	199.8	0.0	282.8
End	. . . (Jupiter is near opposition.)	. . . (Moon is near new.)	. . . (For more accurate value use LONGITUDE.)

Program EGYPTDATE

This program gives the ASTRODATE for any date in the Egyptian Civil Calendar if the year B.C. is known. It is based on the historic statement that Season 1, Month 1, Day 1 of the Egyptian Civil Calendar fell on July 20, 139 A.D. in the Julian, OS, calendar.

The ancient Egyptian calendar contained exactly 365 days, no more, no less, a constant length. It would slip against the modern calendar by about one day every four years. The ancient Egyptian calendar was divided arbitrarily into three seasons, called "Inundation," "Going forth," and "Drought." Because of the ¼-day error, these names did not always fit the flooding of the Nile, and the going forth into the fields, etc. Each of the "seasons" had 4 months of exactly 30 days each. A period of "bad luck," 5 extra days, was added on at the end of season 3. The calendar read as follows:

Season 1, month 1, 2, 3, 4.
Season 2, month 1, 2, 3, 4.
Season 3, month 1, 2, 3, 4.
Year-end, day 1, 2, 3, 4, 5 . . . total number of days 365.

The natural seasons, including the inundation, had different dates year by year in the Civil Calendar, depending on the epoch. The moon's phases were at no fixed day number (as with the Jewish calendar), and most of the cycles of astronomy did not fit this 365-day year. But, unlike our present Gregorian calendar of length 365 97/400 days, the Egyptian system did give a continuous day count, whereas our leap-year system does not.

$$\text{ASTRODATE} = 139.5482 - 365/365.25 \times \text{Int: } \underline{\underline{\text{year B.C.} + 138.5482}/365 \times 365.25}$$

$$+ 1/365.25\,(30 \times \underline{\text{MONTH} - 1} + 120 \times \underline{\text{SEASON} - 1} + \text{DAY} - 1)$$

Relying on the historic data, the program is accurate to within 1 day.

Program	Example
EGYPTDATE	1272 B.C.
	MONTH 2
	SEASON 2
	DAY 19
enter year B.C.	1272
+ 138.5482546	138.5482
=	1410.5482
divide by 365	365
=	3.864
× 365.25	365.25
=	1411.5
integer	1411
× 365	365
=	515015
divide by	
365.25	365.25
=	1410.034
+/−	−1410.034
+139.5482546	139.5482
=	−1270.485
display is	
ASTRODATE for	
Egyptian New	
Year's Day,	
S 1, M 1, D1 . . .	−1270.485
store in M 1	
enter MONTH	2
−1	1
=	1
× 30	30
=	30
store in M 2	
enter SEASON	2
−1	1
=	1
× 120	120
=	120
add to M 2	
enter DAY	19
−1	1
=	18
+	
recall M 2	150
=	168
divide by	
365.25	365.25
=	0.459
+	
recall M 1	−1270.485
=	−1270.026
display is	
ASTRODATE for	
given Egyptian	
date . . .	−1270.026
End	• • •

. . . continue in next column

Program CALENDAR

This program converts a value of ASTRODATE into the date in the Julian Calendar, the Old Style calendar that was in use historically up to the 16th century. The input is ASTRODATE. The output is the YEAR number, either A.D. or B.C., and DAYS, the number of calendar days that have elapsed since January 1.0. The quantity DAYS can be converted to the calendar Month, DAY, and fraction of the DAY where 0.0 is midnight U.T. The fraction can be converted to the time U.T.

If ASTRODATE is negative,

YEAR = Int: (ASTRODATE − 1),

and the year number in the B.C. system is (−YEAR + 1).

If ASTRODATE is positive,

YEAR = Int: ASTRODATE

For positive or negative ASTRODATE,

DAYS = ASTRODATE X 365.25 − 365 × YEAR − Int: YEAR + 4711/4 + 1177

Example 1: ASTRODATE −583.5934292 gives
−584 A.D., i.e., 585 B.C., May 28, 12:00 U.T.

Example 2: ASTRODATE 5.239151266 gives
5 A.D., March 28, 14:24 U.T.

Program	Example 1	Example 2
CALENDAR	−583.5934292	5.239151266
enter . . .		
ASTRODATE	−583.5934292	5.239151266
store in M 1		
is ASTRODATE negative?	yes	no
−1	1	↓
=	−584.593	
integer	−584	5
store in M 2		
+/−	584	−5
+1	1	1
=	585	−4
store M 3		
recall M 1	−583.593	5.239
X 365.25	365.25	365.25
=	−213157.5	1913.6
store M 4		
recall M 2	−584	5
X 365	365	365
=	−213160	1825
+/−	213160	−1825
add to M 4		
recall M 2	−584	5
+4711	4711	4711
=	4127	4716

(continued on next page)

Program	Example 1	Example 2
divide by 4	4	4
=	1031.75	1179.0
integer	1031	1179
+/−	−1031	−1179
add to M 4		
recall M 2	−584	5
display is . . .		
YEAR A.D. . . .	−584	5
is YEAR		
negative?	yes	no
recall M 3	585	↓
display is		
YEAR B.C. . . .	585	
enter 1177	1177	1177
add to M 4		
recall M 4		
display, DAYS . . .	148.5	86.6
from the table of values of A, find the A-value which is closest to, but lower than, DAYS. (Allow for LEAP!) This A-value gives		
MONTH . . .	MAY	MARCH
enter A-value	121	59
+/−	−121	−59
+1	1	1
=	−120	−58
+		
recall M 4	148.5	86.6
=	28.5	28.6
display is		
DAY number . . .	28.5	28.6
fraction	0.5	0.6
× 24	24	24
=	12.0	14.4
display is		
HOUR U.T.	12.0	14.4
fraction	0.0	0.4
× 60	60	60
=	0	24
display is		
MINUTES . . .	0	24
End	• • •	• • •

Program MINDSTEP

This program gives MINDSTEP, the year A.D. corresponding to mindstep number M as used in the empirical equation assumed in Chapter 14. The value of M can be an integer or a decimal number. Caveat: Future mindsteps may not follow this equation.

$$\text{MINDSTEP} = 2053.7 - 32682.7/4^M$$

Program	Example 1	Example 2	Example 3
MINDSTEP	M = 4	M = 3.9	M = 3.0
enter 2053.7	2053.7	2053.7	│
store in M 1			│
enter 32682.7	32682.7	32682.7	│
store in M 2			│
(label 1)	↓	↓	↓
enter value M	4.0	3.9	3.0
store in M 3			
enter 4.0	4.0	4.0	4.0
logn	1.386	1.386	1.386
×			
recall M 3	4.0	3.9	3.0
=	5.545	5.406	4.158
exp(x)	256.0	222.8	64.0
store in M 4			
recall M 2	32682.7	32682.7	32682.7
divide by			
recall M 4	256.0	222.8	64.0
=	127.66	146.65	510.66
+/−	−127.66	−146.65	−510.66
+			
recall M 1	2053.7	2053.7	2053.7
=	1926.0	1907.0	1543.0
display is MINDSTEP	1926 A.D.	1907 A.D.	1543 A.D.
(go back to label 1)	↑	↑	
End	* * *	* * *	* * *

Program STONEHENGE

This program gives the declination error, delta d, and the vertical error, V, in a suspected archaeoastronomical alignment. The output V is the angle that the chosen target is above the skyline in the direction indicated by the azimuth of the line in the ancient structure. Three types of targets are programmed—a point object such as a star, a planet or the center of a disk, and a disk object under the two conditions of tangent standing like a wheel on the skyline, or tangent below the skyline in the condition called "first flash."

Definitions:

r	the angle of refraction caused by the earth's atmosphere
S	the altitude in degrees of the distant skyline above the horizontal plane
f	the correction factor for site elevation above mean sea level measured in meters

d	the astronomical declination of that part of the target that is exactly on the skyline
G	the geometric altitude of the skyline as seen from a zero-diameter, airless earth
A	the azimuth of the archaeo-line measured East of North
L	the latitude of the site
p	the parallactic displacement in target produced by length of one earth radius
delta d	the difference in declination between chosen target and the declination on the skyline
d_{obj}	the declination of the chosen target object (center of disk for sun or moon)
q	one half the diameter of the disk in degrees
B	the slant angle made by the rising or setting object as it moves near the skyline
V	the vertical error between the chosen target and the skyline in the direction of the alignment
H	the elevation of the site above mean sea level measured in meters

Values:

q = 0 for a star or planet
q = 0 for the center of the disk of sun or moon
q = −0.267 for sun tangent on horizon in Northern Hemisphere
q = +0.267 for sun tangent below horizon in Northern Hemisphere
q = −0.259 for moon tangent on horizon in Northern Hemisphere
q = +0.259 for moon tangent below horizon in Northern Hemisphere
p = 0.002 for sun
p = 0.951 for moon

Sun and Moon Extreme Declination

Year, B.C.	Sun	High Moon	Low Moon
4000	24.11	29.26	18.96
3000	24.03	29.18	18.88
2000	23.93	29.08	18.78
1000	23.82	28.97	18.67

The algorithm:

$r = 0.58\ f \exp(-S/3.0)$
$f = \exp(-H/8400)$
$\sin d = \sin L \sin G + \cos L \cos G \cos A$
$G = S - r + p$
$\text{delta } d = d_{obj} + q \cos B - d$
$\sin B = \cos L \sin A / \cos d_{obj}$
$V = \text{delta } d \sec B \times (1 - r/3.0)$

Example: sunrise over the horizon at Stonehenge in the direction of the Heel stone.

Input		stored in	Output		stored in
L	51.178	M 1	V	0.4486406482	display
d_{obj}	23.906	M 2	d	23.23091814	M 12
q	−0.267	M 3	delta d	0.449	M 6
p	0.002	M 4	G	0.132867499	M 0
H	102	M 5	B	32.31550039	M 8
A	51.222	M 6	r	0.469132501	M 9
S	0.6	M 7			

(continued on next page)

Program	Example
STONEHENGE	
enter L	51.178
store in M 1	
enter d_{obj}	23.906
store in M 2	
enter q	−0.267
store in M 3	
enter p	0.002
store in M 4	
enter H	102
store in M 5	
enter A	51.222
store in M 6	
enter S	0.6
store in M 7	
recall M 6	51.222
sine	0.779
store M 11	
recall M 1	51.178
cosine	0.626
×	
recall M 11	0.779
=	0.488
store M 11	
recall M 2	23.906
cosine	0.914
divide by	
recall M 11	0.488
=	1.870
1/x (reciprocal)	0.534
arcsine	32.315
store M 8	
recall M 7	0.6
divide by 3.0	3.0
=	0.2
+/−	−0.2
exp(x)	0.818
× 0.58	0.58
=	0.474
store M 11	
recall M 5	102
divide by 8400	8400
=	0.012

. . . continue in next column

Program	Example
+/−	−0.012
exp(x)	0.987
store M 12	
×	
recall M 11	0.474
=	0.469
store M 9	
recall M 7	0.6
−	
recall M 9	0.469
=	0.130
+	
recall M 4	0.002
=	0.132
store M 0	
sine	0.0023
store M 11	
recall M 1	51.178
sine	0.779
×	
recall M 11	0.0023
=	0.0018
store M 11	
recall M 6	51.222
cosine	0.626
store M 12	
recall M 0	0.132
cosine	0.999
×	
recall M 12	0.626
=	0.626
store M 12	
recall M 1	51.178
cosine	0.626
×	
recall M 12	0.626
=	0.392
store M 12	
recall M 11	0.0018
add to M 12	
recall M 12	0.394
arcsine	23.230
store M 12	

. . . continue in next column

Program	Example
recall M 8	32.315
cosine	0.845
×	
recall M 3	−0.267
=	−0.225
+	
recall M 2	23.906
=	23.680
−	
recall M 12	23.230
=	0.449
store in M 6	
recall M 8	32.315
cosine	0.845
1/x	1.183
×	
recall M 6	0.449
=	0.531
store M 11	
recall M 9	0.469
divide by 3.0	3.0
=	0.156
+/−	−0.156
×	
recall M 11	0.531
=	−0.083
+	
recall M 11	0.531
=	0.4486
display is	
V, vertical	
error	0.4486
recall M 12	23.2309
display is	
d, declination	
on skyline . . .	23.2309
recall M 6	0.449
display is	
delta d,	
error in	
declination . . .	0.449
End	• • •

Program LONGITUDE

This program gives the heliocentric coordinates of the earth and 5 planets, and it gives the geocentric LONGITUDE of the sun, moon, moon's ascending mode, and 5 planets. Only the algorithm is given below, but the program steps required to place it on a home computer, although long and tedious, are straightforward, and the author is confident that by this stage the reader will be able to complete the operation. In principle there is nothing to stop LONGITUDE being computed on a nonprogrammable pocket calculator, but the process will be completed more efficiently on a programmable home computer with subroutines and generous memory storage.

Subscripts:

1 MERCURY	6 SATURN
2 VENUS	7 MOON
3 EARTH	8 MOON'S ASCENDING NODE
4 MARS	0 SUN
5 JUPITER	

The equations assume that a planet moves around an elliptical orbit with the sun at the focus at a distance r. The position is given by angle M, measured from the line of the major axis, starting at perihelion. The equations also assume that the moon and the sun each move in an elliptical orbit with the earth at the focus and M measured from the major axis, starting at perigee for the moon and perihelion for the sun.

$$M = 360 \times \text{Pos. F} : \underline{\text{ASTRODATE} - \text{STARTP/PERIODP}}$$

$$r = D \times \text{reciprocal of: } \underline{1 + e \times \cos(M + E \times \sin M)}$$

The angle in the orbit measured from the vernal equinox, θ, is

$$\theta = 360 \times \text{Pos. F} : \underline{\text{ASTRODATE} - \text{START/PERIOD}} + E \times \sin M + H \times (\text{ASTRODATE} - 1000)^2$$

The Cartesian coordinates are given by

$x = r \times \cos\theta$, and $y = r \times \sin\theta$

Heliocentric Coordinates

For planets 1,2,4,5, and 6, the heliocentric coordinates are (r_1, θ_1) and (x_1, y_1).

For the earth, the heliocentric coordinates are given by

$r_3 = -r_0 \quad x_3 = -x_0 \quad \text{and} \quad y_3 = -y_0$

Heliocentric coordinates are not required for the moon.

Geocentric Coordinates

These coordinates are the geocentric distance R, and LONGITUDE.

For the sun and moon,

$R_0 = r_0,\ R_7 = r_7$

$\text{LONGITUDE}_0 = \theta_0$; $\text{LONGITUDE}_7 = \theta_7$; and $\text{LONGITUDE}_8 = \theta_8$

(continued on next page)

For planets 1,2,4,5, and 6, the geocentric Cartesian coordinates are

$$XG_1 = x_1 - x_3;\ YG_1 = y_1 - y_3;$$

and the geocentric polar coordinates are

$$R_1 = \sqrt{XG_1^2 + YG_1^2}$$

$$\text{LONGITUDE}_1 = \text{arctangent}\ (YG_1/XG_1),$$

but . . . if LONGITUDE is a negative angle, $-L$, then

$$\text{LONGITUDE}_1 = -L + 360$$

(Note that many calculators and computers have a subroutine key for coordinate conversion.)

The values of r, x, y, XG, YG and R are given in astronomical units. To convert to miles, R(miles) = 92,956,000 X R and R(kilometers) = 149,598,000 X R.

	Example 1		Example 2	
	1484 B.C. Aug. 16.5 OS ASTRODATE −1482.375086		1982 A.D. March 10.0 NS ASTRODATE 1982.151951	
	LONGITUDE	R	LONGITUDE	R
Sun	129.3	1.000	349.1	0.992
Moon	275.8	0.0024	172.4	0.0025
Mercury	148.7	0.721	324.4	1.117
Venus	144.7	0.288	305.0	0.504
Mars	129.3	2.631	196.2	0.702
Jupiter	138.4	6.432	219.8	4.749
Saturn	273.8	8.868	201.9	8.831
Moon node	160.8	* * *	109.5	* * *

Coefficients

	1 MERCURY	2 VENUS	4 MARS	5 JUPITER
STARTP	999.852286	999.568562	999.229948	990.968758
PERIODP	0.24084695	0.61519759	1.88089230	11.86275688
START	999.810861	999.367887	999.451195	991.029042
PERIOD	0.24084446	0.61518264	1.88071166	11.85654247
E	23.42	0.83	10.59	5.37
H	3×10^{-8}	3×10^{-8}	3×10^{-8}	2×10^{-8}
e	0.205428	0.007249	0.092497	0.046867
D	0.370763	0.723292	1.510643	5.191174

	6 SATURN	7 MOON	8 MOON'S ASCENDING NODE	0 SUN
STARTP	972.007819	1899.903726	0	999.946301
PERIODP	29.47070046	$7.544024898 \times 10^{-2}$	1	1.00002630
START	995.468903	1899.909577	1913.36620	999.208019
PERIOD	29.42376925	$7.480241514 \times 10^{-2}$	−18.612904	0.99997879
E	6.76	6.29	0	1.96
H	5×10^{-8}	1.133×10^{-7}	2.078×10^{-7}	3×10^{-8}
e	0.058984	0.054900	0	0.01713
D	9.521689	0.00256177	0	0.999708

Accuracy: Mercury, Moon, 5 degrees; other planets, 3 degrees; sun, 0.3 degrees.

Bibliography

Aveni, Anthony. *Archaeoastronomy in pre-Columbian America.* Austin: University of Texas Press, 1975.

Berendzen, R.; Hart, R.; and Seeley, D. *Man Discovers the Galaxies.* New York: Science History Publications, 1976.

Breuil, H., and Obermaier, H. *La Pileta.* Monaco; 1915.

Budge, E. A. Wallis. *Egyptian Magic.* New York: Dover, 1948.

Burl, Aubrey. *Rites of the Gods.* London; J. M. Dent, 1981.

______. *Prehistoric Averbury.* New Haven: Yale University Press, 1979.

Cornell, James. *The First Stargazers.* New York: Charles Scribner's Sons, 1981.

Daniel, Glyn. *The Idea of Prehistory.* New York: Pelican, 1964.

Duffet-Smith, Peter. *Practical Astronomy with your Calculator.* England: Cambridge University Press, 1979.

Eccles, John. *The Human Mystery.* New York: Springer-Verlag, 1979.

Edwards, I. E. S. *The Pyramids of Egypt.* New York: Pelican, 1961.

Edwards, I. E. S.; Gadd, C. J.; and Hammond, N. G. L. *The Cambridge Ancient History,* vol. 1, part 1. England: Cambridge University Press, 1970.

Fraser, J. G. *The Golden Bough.* London, Macmillan, 1911.

Hadingham, Evan. *Secrets of the Ice Age.* London: Heinemann, 1979.

______. *Circles and Standing Stones.* New York: Doubleday, 1976.

Ledrain, E. *Les Monuments Egyptiens.* Bibliothèque de l'école des hautes Etudes, vol. 47, 1881.

Marshack, Alexander. *Science,* vol. 146, 1964, p. 146.

McClintock, Marshall. *Prescott's "The Conquest of Mexico."* New York: Julian Messner, 1948.

McMann, Jean. *Riddles of the Stone Age.* London: Thames & Hudson, 1980.

Mendelssohn, Kurt. *The Riddle of the Pyramids.* New York: Praeger, 1974.

Ono Yoshimasa, A. "How I (Einstein) Created the Theory of Relativity." *Physics Today,* August 1982.

Ornstein, Leonard. "A Biologist Looks at the Numbers." *Physics Today,* March 1982.

Osborne, Harold. *South American Mythology.* England: Paul Hamlyn, 1968.

Pritchard, James B. *Ancient Near East Texts.* Princeton, N.J.: Princeton University Press, 1955.

Protheroe, W. M.; Capriotti, E. R.; and Newsom, G. H. *Exploring the Universe.* New York: Merrill, 1981.

Sagan, Carl. *Cosmos.* New York: Random House, 1980.

Scott, Joseph and Lenore. *Egyptian Hieroglyphs for Everyone.* New York: Funk and Wagnalls, 1968.

Shackley, Myra. *Neanderthal Man.* London: Duckworth, 1980.

Thom, Alexander. *Megalithic Lunar Observatories.* England: Oxford University Press, 1971.

Watson, Lyall. *Lifetide: The Biology of Consciousness.* New York: Simon and Schuster, 1980.

Index

Telepathy, 308
Telescope, 183, 188, 194-201, 203, 222

ILLUSTRATION CREDITS

Gerald Hawkins: pp. 8, 9, 11, 13, 20, 21, 34 (bottom right), 36, 47, 60–62, 63, 71–74, 76, 78–80, 86, 91, 92 (top), 98–100, 102, 103, 105, 108, 115, 116, 120, 122, 124, 125, 127–131, 135, 139, 141, 150, 152, 154 (top), 160, 161, 166, 170, 174, 175, 178, 182, 184, 187, 190, 192, 197, 200, 215, 241, 278, 295, 297, 298; Deri Barringer: p. 18; Staatliche Mussen, Berlin DDR: p. 29; Hale Observatories: pp. 34 (top), 35, 199, 202, 204, 210, 211, 218, 220, 224, 227, 228, 232, 233, 237, 239, 240, 242; Tony Morrison: p. 34 (bottom left); Carmen Cook de Leonard: pp. 56–59; Richard Brinckerhoff: p. 84; Michael J. O'Kelly: p. 92 (bottom); James Cornell: p. 119; The Louvre, Paris: p. 134; Otto Neugebauer: p. 154 (bottom); Biblioteca Ambrosiana, Milan: p. 168; Biblioteca Nazionale, Firenze: p. 189; U.S. Naval Observatory: p. 205; U.S. Naval Research Laboratory: p. 212; Association of Universities for Research in Astronomy, Inc.: p. 225; NASA: pp. 252, 256–260, 262, 264–267, 269, 270, 272, 274, 281, 290–292; U.S. National Radio Observatory: pp. 280, 287.

ILLUSTRATION CREDITS

SERIES ON KNOTS AND EVERYTHING

Editor-in-charge: Louis H. Kauffman *(Univ. of Illinois, Chicago)*

The Series on Knots and Everything: is a book series polarized around the theory of knots. Volume 1 in the series is Louis H Kauffman's Knots and Physics.

One purpose of this series is to continue the exploration of many of the themes indicated in Volume 1. These themes reach out beyond knot theory into physics, mathematics, logic, linguistics, philosophy, biology and practical experience. All of these outreaches have relations with knot theory when knot theory is regarded as a pivot or meeting place for apparently separate ideas. Knots act as such a pivotal place. We do not fully understand why this is so. The series represents stages in the exploration of this nexus.

Details of the titles in this series to date give a picture of the enterprise.

Published:

Vol. 1: Knots and Physics (3rd Edition)
L. H. Kauffman

Vol. 2: How Surfaces Intersect in Space — An Introduction to Topology (2nd Edition)
J. S. Carter

Vol. 3: Quantum Topology
edited by L. H. Kauffman & R. A. Baadhio

Vol. 4: Gauge Fields, Knots and Gravity
J. Baez & J. P. Muniain

Vol. 5: Gems, Computers and Attractors for 3-Manifolds
S. Lins

Vol. 6: Knots and Applications
edited by L. H. Kauffman

Vol. 7: Random Knotting and Linking
edited by K. C. Millett & D. W. Sumners

Vol. 8: Symmetric Bends: How to Join Two Lengths of Cord
R. E. Miles

Vol. 9: Combinatorial Physics
T. Bastin & C. W. Kilmister

Vol. 10: Nonstandard Logics and Nonstandard Metrics in Physics
W. M. Honig

Vol. 11: History and Science of Knots
edited by J. C. Turner & P. van de Griend

Vol. 12: Relativistic Reality: A Modern View
edited by J. D. Edmonds, Jr.

Vol. 13: Entropic Spacetime Theory
J. Armel

Vol. 14: Diamond — A Paradox Logic
N. S. Hellerstein

Vol. 15: Lectures at KNOTS '96
S. Suzuki

Vol. 16: Delta — A Paradox Logic
N. S. Hellerstein

Vol. 19: Ideal Knots
A. Stasiak, V. Katritch & L. H. Kauffman

Vol. 20: The Mystery of Knots — Computer Programming for Knot Tabulation
C. N. Aneziris

Vol. 24: Knots in HELLAS '98 — Proceedings of the International Conference on Knot Theory and Its Ramifications
edited by C. McA Gordon, V. F. R. Jones, L. Kauffman, S. Lambropoulou & J. H. Przytycki

Vol. 26: Functorial Knot Theory — Categories of Tangles, Coherence, Categorical Deformations, and Topological Invariants
by David N. Yetter

Vol. 27: Bit-String Physics: A Finite and Discrete Approach to Natural Philosophy
by H. Pierre Noyes; edited by J. C. van den Berg

Vol. 29: Quantum Invariants — A Study of Knots, 3-Manifolds, and Their Sets
by Tomotoda Ohtsuki

Vol. 30: Symmetry, Ornament and Modularity
by Slavik Vlado Jablan

Vol. 31: Mindsteps to the Cosmos
by Gerald S Hawkins